Just Enough Algebra
for Students of Statistics

Ronald I. Brent
University of Massachusetts–Lowell

Guntram Mueller
University of Massachusetts–Lowell

ADDISON-WESLEY

An imprint of Addison Wesley Longman, Inc.

Reading, Massachusetts • Menlo Park, California • New York • Harlow, England
Don Mills, Ontario • Sydney • Mexico City • Madrid • Amsterdam

Library of Congress Cataloging-in-Publication Data

Brent, Ronald I.
    Just-enough algebra for students of statistics / Ronald I.
Brent, Guntram Mueller.
      p. cm.
    Includes index.
    ISBN 0-201-50344-1
    1. Algebra. I. Mueller, Guntram. II. Title.
    QA152.2 .B74 1999
    512--ddc21

                                                    98-45578
                                                        CIP

Reproduced by Addison Wesley Longman from camera-ready copy supplied by the authors.

1 2 3 4 5 6 7 8 9 10 CRS 01009998

Table of Contents

To the Student

Flipping through the pages of your Statistics textbook, you'll see a lot of equations and algebraic expressions that may remind you of the algebra you studied. You'll also see a lot of symbols that may be new to you. Symbols like Σ, and some x's and f's that have odd little characters on top like $-$, $\sim$ and $\wedge$ (officially known as "bar", "tilde", and "hat"). As you think back on your algebra experience, you may wonder: "How much algebra do I really need to know? And how well do I need to know it?" Well, as they say, there's good news and there's bad news. The good news: You don't need to know very much algebra at all. And now the bad news: The parts you do need, however, you've got to know very well. You want your algebra skills to be good enough so you can depend on them without having to think about them too much, leaving your mind open and available for thinking about what you are really studying: Statistics!

So what do you need to know? Well, you need to know how to handle fractions, decimals, and percents, and also exponents and square roots. You need to know how to simplify and evaluate algebraic expressions, and how to work with equations and inequalities. Finally, you have to interpret graphs and even create your own graphs.

The idea of this book is simply to give you the algebra you need to succeed at Statistics. In reading the Table of Contents, however, you may notice that there are a few other things that you don't recognize from your algebra classes. No, that's not something you missed when you were out sick with the flu during Algebra II. In fact, it's not algebra at all, but we added it to this book because many Statistics books don't take a lot of time on these topics, on the assumption that you've seen them before. We believe you'll be glad to have the material in Chapters Three and Four, on summation, factorials, permutations, and combinations. We have also included many examples illustrating how to use the TI-83, as well as other scientific calculators, to solve basic statistical problems.

What's the best way to use this book? Here's how.

a) The ideal way: Before your statistics course even starts, use this book to brush up on your algebra skills, retouching the things you may have partly forgotten, and clearing up some misconceptions you may have had. Read one section at a time, making sure it makes sense to you. To test yourself, do the exercises, especially the latter half of each assignment.

b) If it's too late for that: As soon as you can, set aside a chunk of time to go through Chapters One and Two. If you can do the harder looking exercises correctly, that should do it. While doing this, however, it is absolutely essential that you also keep up-to-date with your statistics homework. Now, as for the other chapters, you may want to use them as the need arises. If inequalities confuse you, look at Chapter One and Chapter Seven. If it's those mysterious Σ's, there's Chapter Three to help you.

Best of luck to you! You will be learning one of the most useful subjects in mathematics, seeing how to extract meaning from incomplete information. Maybe more importantly, you'll see how to recognize the strengths and weaknesses of these methods and the degree of validity to ascribe to them.

To the Instructor

The difficulties that students have in learning mathematics are compounded by the fact that mathematical subjects build upon one another. Those of us who teach statistics are familiar with this problem. We see many statistics students who stumble over the algebraic stepping stones, or get confused using the calculator, and then lose sight of the statistics to be learned. The actual amount of algebra needed to succeed at statistics is not that great, yet it does need to be mastered.

This book is intended for students in statistics courses for health, humanities, social sciences, education, and business or economics majors. It gives a clean and concise review of the algebra needed to study statistics. Chapters One and Two are a quick presentation of all the basics of arithmetic and associated manipulative tasks, as well as various mathematical notations and manipulation of algebraic expressions. We have included Chapter Three on subscripts and summation, mean and standard deviation, as well as Chapter Four on factorials, permutations, and combinations, so as to give students a more detailed look at these subjects; they are often given a cursory treatment in statistics books, as they hurry on to other things because these subjects are assumed, rightly or wrongly, to have been covered already. Our presentation of graphs in Chapter Five pays close attention to the various representations of data, and shows how tables and graphs and sometimes even algebraic expressions can be changed from one to the other. Chapter Six contains a soup-to-nuts description of lines, which is good to review the day before you begin linear regression. Finally, in Chapter Seven we cover the solution and interpretation of the various equations and inequalities that statistics students will encounter in their studies.

A key feature of this book is that we have also included many examples illustrating how to use the TI-83, as well as other scientific calculators, to solve basic statistical problems. These include the simple problem of determining a sample mean and standard deviation, as well as the more complicated problems of graphing histograms, scatter diagrams, and regression lines.

Acknowledgments

We wish to acknowledge all those who helped in this venture, beginning with our wives, Leor and Edie, and our children, Sarah and Adam, and Ariadne. Thanks to the reviewers: Mario Triola, Tommy Leavelle, Bill Meisel, Chris Ennis, and Chris Burditt, for all their excellent suggestions that helped to create the final product. We especially wish to thank Laurie Rosatone, Ellen Keohane, and Kim Ellwood at Addison Wesley Longman.

University of Massachusetts Lowell   Ronald I. Brent

Guntram Mueller

Chapter 1

Numbers and Their Disguises

Every number can be written in many different forms. For example, the numbers $\frac{8}{12}$, $\frac{10}{15}$, $\dfrac{3}{4 + \dfrac{9}{18}}$, and even $0.6666\cdots$, are all really just different ways of writing the number $\frac{2}{3}$. (Check it out; don't take <u>our</u> word for it.) For most purposes, the idea is to keep things as simple as possible, and just to use the form $\frac{2}{3}$.

Sometimes it is better to have a mathematical expression written as a product, other times it is better to have a sum; it all depends on what you need to do with it. In any event, <u>changing the form of an expression is something you have to do all the time</u>. Correctly! It's the nuts and bolts of mathematics, and comes up in any quantitative work.

1.1 <u>Rounding Decimals</u>

Some decimal expressions for numbers are straightforward and neat, like $\frac{2}{5} = 0.4$ and $\frac{347}{10,000} = 0.0347$, but then there are others, like $\frac{2}{3}$, which are neither straightforward nor neat. We write $\frac{2}{3} = 0.6666\cdots$ where the " $\cdots$ " means to keep on writing 6's, endlessly. (Sounds exhausting.) Here's the good news: Whenever the exact decimal expression of a number is so long that it's awkward to use, you can "round it off" using a smaller number of digits.

How to round off a number:

a)   Make the "cut" at the desired place.

b)   If the digit after the cut is 5 or greater, increase the digit before the cut by 1. Otherwise, leave it as is.

> **Example 1:**   Round off 43.8742189 and -1.32505 to two decimal places.
>
> **Solution:** $43.8742189 = 43.87$
>
> and $-1.32505 = -1.33$
>
> Notice that the answer is -1.33 and not -1.32, because the next digit after the 2 was 5. ■

The result of rounding is an approximation to the actual number in question, and the more digits we keep, the better the approximation. How good an approximation is needed? That depends upon the particular situation. In statistics, a good rule-of-thumb is to <u>round off using one more decimal place than the original data</u> that went into the calculation. When you're using a calculator, never use the eight or ten digits given in the answer. Always round off the answer to the number of digits that is appropriate for your purposes.

> **Example 2:**   Your test scores are 80, 65, and 82. What is your average score?
>
> **Solution:**   The average test score is $\dfrac{80 + 65 + 82}{3}$ which equals
>
> $\dfrac{227}{3} = 75.6666\cdots$, which can be rounded to 75.7, right?
>
> Notice the extra decimal place, and notice that the 6 following the decimal had to be rounded up to 7. ■

Exercises 1.1

1)   Round off each of the following numbers to one decimal place.

   a)   2.875

   b)   −10.3283

   c)   153.8

   d)   0.0983

   e)   0.00983

2)   Round off each of the following numbers to two decimal places.

   a)   4.7321

   b)   −0.02934

   c)   1.446

   d)   −143.143143

3)   If your test scores are 93, 81, 65, and 76, what is your average? Use the rule of thumb given above to round your answer.

4)   Suppose the temperature at Club Arctic is taken at 1 PM everyday for a week. The seven temperatures, measured in Centigrade, are given as:

Sun.	Mon.	Tues.	Wed.	Thurs.	Fri.	Sat.
1.4°	−2.6°	0.3°,	1.2°,	−1.1°,	1.8°,	1.5°

What was the average temperature at 1 PM that week?

1.2 Brackets

Brackets are ways of "packaging" or grouping numbers together.

Example 1:

$$5 - \left(1 + \frac{1}{2}\right) + (3 - 4) - \left(7 - \frac{1}{2}\right)$$

$$= 5 - (1\tfrac{1}{2}) + (-1) - (6\tfrac{1}{2})$$

$$= 5 - \frac{3}{2} - 1 - \frac{13}{2}$$

$$= 5 - 1 - \frac{3}{2} - \frac{13}{2}$$

$$= 4 - \frac{16}{2} \;=\; 4 - 8 \;=\; -4 \quad \blacksquare$$

You may prefer to get rid of the brackets. Here's how: a) if there's a "+" in front, leave all the signs of the terms inside as they are; b) if there is a "−" in front, change all the signs of the terms inside.

Example 2:

$$5 - \left(1 + \frac{1}{2}\right) + (3 - 4) - \left(7 - \frac{1}{2}\right)$$

$$= 5 - 1 - \frac{1}{2} + 3 - 4 - 7 + \frac{1}{2} \quad \text{(Why is it } +\frac{1}{2} \text{ ?)}$$

$$= 5 - 1 + 3 - 4 - 7 - \frac{1}{2} + \frac{1}{2}$$

$$= -4 \quad \blacksquare$$

Example 3:   Simplify: $3 - 2(8 + 1) - 3(5 - 7)$

Solution: Method 1: $3 - 2(9) - 3(-2) = 3 - 18 + 6 = -9$

Method 2: $3 - 16 - 2 - 15 + 21 = -9$ ∎

Notice that in the second method in Example 3, the 8 and 1 were both multiplied by -2, and the 5 and -7 were multiplied by -3. This is based on the property of distributivity – namely, $a(b + c) = ab + ac$. Method 1 is easier in this case, but in many other cases Method 2 will be needed. <u>You've got to know both</u>!

You will notice that your calculator has a set of bracket or parenthesis keys. They look like ⬚ (and ⬚) . You will sometimes have to use those keys to make sure that you are calculating the correct number. Consider the following example.

Example 4:   Compute $1.37 - (4.3 - 2.8)$ on your calculator using the parenthesis keys.

Solution: This calculation would be handled on the TI-83 as :

1.37 [−][( ] 4.3 [−] 2.8 [ )]   [ENTER]

The answer is $-.13$, as seen on the screen.

```
1.37-(4.3-2.8)
            -.13
■
```

Remarks: 1) For the remainder of the book, calculator steps will be first given for the TI-83. Most of the calculations will be identical, but where needed, differences in other scientific calculator usage will be pointed out.

2) The $\boxed{\text{ENTER}}$ key on the TI-83 is replaced by the $\boxed{=}$ key on most other scientific calculators.

3) For now, we have used the $\boxed{\text{MODE}}$ key to set the mode to Normal, instead of Sci or Eng. This gives numbers in decimal form. You may want to do this so your calculator screen will look just like ours! On the other hand, if you use Sci or Eng, the answer may be given in scientific notation. (See Section 2.2.)

In algebra, letters stand for numbers, so the laws of arithmetic apply to them in exactly the same way:

Example 5: $x y - (2x - y) - 2y(1 - x)$

$$= xy - 2x + y - 2y + 2yx$$

$$= 3xy - 2x - y \quad \blacksquare$$

Notice that xy and $2yx$ add up to $3xy$.

Example 6:

$$3x^2y - (x^2 - y^3) - 2y(x - y)$$

$$= 3x^2y - x^2 + y^3 - 2xy + 2y^2 \quad \blacksquare$$

Notice the "+" in front of the y in Example 5 and the y^3 in Example 6, as well as the use of the distributive law in both examples. If the exponents in the last example are confusing you, skip ahead to Section 2.1 and then return here.

Exercises 1.2

1)   Simplify: $4 - (5 - 3) + 3(4 - 7) - 4(1 + 2)$

2)   Simplify: $-2(3 - 4) + 4(5 + 3) - 3(2 - 6)$

3)   Simplify: $5(8 - 4) - 3(6 + 4) - 3(28 - 16)$

4)   Using your calculator's parenthesis keys, compute $1.83 - (4.37 - 15.85)$.

5)   Using your calculator's parenthesis keys, compute $2.8 + (-4.8 + 11.3)$. Do you really need to use the parenthesis keys in this case?

6)   Using your calculator's parenthesis keys, compute the expression
$$1.3(2.8 - 1.9) + 0.5(10.1 - 4.8).$$

In Exercises 7-10, simplify the given expressions.

7) $4xy - (x - 2xy) - 2y(x - 1)$

8) $(s - t) - (u - t) - (v - u) - (s - v)$

9) $2x(y - 3) - y(x + xy) + 2y(x + 1)$

10) $x(y + z) - z(x + y) + 2y(x - z) - x(3y - 2z)$

1.3 <u>Multiplying and Dividing Fractions</u> (We'll do adding and subtracting later.)

Multiplying fractions is the easiest manipulative task. The numerator is the product of the given numerators, and the denominator is the product of the given denominators. That is

$$\frac{a}{b} \cdot \frac{c}{d} = \frac{a \cdot c}{b \cdot d} = \frac{ac}{bd} .$$

We use the symbol "·" to denote multiplication; in fact, usually we will omit the symbol altogether.

Example 1:

a) $\dfrac{2}{3} \cdot \dfrac{5}{7} = \dfrac{2 \cdot 5}{3 \cdot 7} = \dfrac{10}{21}$

b) $\dfrac{1}{9} \cdot \left(-\dfrac{5}{8}\right) = \dfrac{1}{9} \cdot \dfrac{-5}{8} = \dfrac{1 \cdot (-5)}{9 \cdot 8} = \dfrac{-5}{72}$

Notice that $-\dfrac{5}{8} = \dfrac{-5}{8}$. Right?   ∎

This multiplication rule can be extended to multiplying more than two fractions simply by multiplying across all the numerators and denominators.

Example 2:

a) $\dfrac{1}{4} \cdot \dfrac{7}{5} \cdot \dfrac{3}{8} = \dfrac{1 \cdot 7 \cdot 3}{4 \cdot 5 \cdot 8} = \dfrac{21}{160}$

b) $\dfrac{-1}{7} \cdot \dfrac{3}{8} \cdot \dfrac{-2}{\pi} = \dfrac{(-1)(3)(-2)}{7 \cdot 8 \cdot \pi} = \dfrac{6}{56\pi} = \dfrac{3}{28\pi}$ ∎

Of course, multiplying a number by 1 produces the same number, so:

$\frac{5}{6} \cdot 1 = \frac{5}{6} \cdot \frac{4}{4} = \frac{20}{24}$. Going backward is usually more important: $\frac{20}{24} = \frac{5 \cdot 4}{6 \cdot 4} = \frac{5 \cdot \cancel{4}}{6 \cdot \cancel{4}} = \frac{5}{6}$.

Here's the key point: you can cancel the factor 4 in the numerator and the denominator. More generally, if a number $c \neq 0$ is a factor of both the top and bottom of a fraction, it may be canceled. When all those common factors are canceled, the fraction is said to be in <u>lowest terms</u>.

Example 3:   Put $\frac{30}{84}$ into lowest terms.

Solution: First factor the numerator and denominator as much as possible and then cancel all common factors.

$$\frac{30}{84} = \frac{\cancel{2} \cdot \cancel{3} \cdot 5}{2 \cdot \cancel{2} \cdot \cancel{3} \cdot 7} = \frac{5}{2 \cdot 7} = \frac{5}{14} \quad \blacksquare$$

Warning: <u>Make sure you cancel only those numbers that are factors of the entire top and the entire bottom.</u> Don't be tempted to try "creative canceling." Consider the expression

$$\frac{3(8) + 7(5)}{3}.$$

Can you cancel the 3's ? NO NO NO NO NO NO NO NO NO !!!! Get

the picture? The problem is that 3 is not a factor of the <u>entire</u> numerator, only of its first term. Hence you cannot simply cancel the 3's. If the expression is

$$\frac{3(8) + 3(5)}{3},$$

however, the situation is different. In this case, <u>each term of the numerator</u> contains the factor of 3 so we may write the numerator as $3(8 + 5)$. Now, that factor of 3 may be canceled with the 3 in the denominator:

$$\frac{3(8) + 3(5)}{3} = \frac{3(8 + 5)}{3} = \frac{\cancel{3}}{\cancel{3}} \cdot \frac{(8+5)}{1} = 1 \cdot \frac{(8+5)}{1} = 13$$

Example 4:   Simplify $\dfrac{3 + 6x}{3}$.

Solution: $\dfrac{3 + 6x}{3} = \dfrac{3(1 + 2x)}{3} = 1 + 2x$ ■

Dividing by a fraction is done by inverting <u>that</u> fraction and multiplying:

$$\frac{\dfrac{a}{b}}{\dfrac{c}{d}} = \frac{a}{b} \cdot \frac{d}{c} = \frac{a \cdot d}{b \cdot c} = \frac{ad}{bc}$$

For example,

$$\frac{\dfrac{-1}{3}}{\dfrac{5}{6}} = \frac{-1}{3} \cdot \frac{6}{5} = \frac{-6}{15} = \frac{-2}{5},$$

or you can cancel early:

$$\frac{-1}{1\cancel{3}} \cdot \frac{\cancel{6}^{2}}{5} = \frac{-2}{5}$$

Example 5: Simplify the following expressions:

a) $\dfrac{\dfrac{2}{3}}{\dfrac{3}{8}}$

b) $\dfrac{\dfrac{5}{4}}{\dfrac{-10}{3}}$

Solution: a) $\dfrac{\dfrac{2}{3}}{\dfrac{3}{8}} = \dfrac{2}{3} \cdot \dfrac{8}{3} = \dfrac{16}{9}$

b) $\dfrac{\dfrac{5}{4}}{\dfrac{-10}{3}} = \dfrac{5}{4} \cdot \dfrac{3}{-10} = \dfrac{\overset{1}{\cancel{5}}}{4} \cdot \dfrac{3}{\underset{-2}{\cancel{-10}}} = -\dfrac{3}{8}$ ∎

Example 6:   Use your calculator to find $\dfrac{\dfrac{13.5}{0.8}}{\dfrac{4.2}{1.1}}$

rounded to two decimal places.

Solution: It's easier to first rewrite this fraction as

$$\dfrac{\dfrac{13.5}{0.8}}{\dfrac{4.2}{1.1}} = \dfrac{13.5}{0.8} \cdot \dfrac{1.1}{4.2}.$$

What is the order in which we need to carry out the operations? Since the above expression is equal to

$$\dfrac{\dfrac{13.5}{0.8} \cdot 1.1}{4.2}$$

you can simply press

13.5 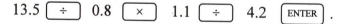 0.8 ✕ 1.1 ÷ 4.2 ENTER .

You will see:

$$\boxed{\begin{array}{r} \texttt{13.5/0.8*1.1/4.2} \\ \texttt{4.419642857} \end{array}}$$

So 4.42 is the answer, rounded to two decimal places. ■

Notice that the TI-83 carries out the operations of multiplication and division in order **as they appear**, NOT like you learned in high school: M $_y$ D $_{ear}$ A $_{unt}$ S $_{ally}$. First the division, then the multiplication, then the second division. See your calculator's manual for a discussion of the order of arithmetic operations. When evaluating complicated expressions you must be clear on the order of operations. See Example 2 in the next section for a case that also includes subtraction. **When in doubt**, you can always use the parenthesis keys to make it all clear. In this case you would use:

You may find yourself having to add or subtract fractional expressions involving algebraic expressions. The game is the same, just remember that the letters merely stand for numbers.

Example 7: Simplify the expression $\left(\dfrac{x^2}{y}\right)\cdot\left(\dfrac{-y+1}{x}\right)$.

Solution: $\left(\dfrac{x^2}{y}\right)\cdot\left(\dfrac{-y+1}{x}\right) = \dfrac{(x^2)\cdot(-y+1)}{y\cdot x}$

$$= \dfrac{x(-y+1)}{y} \text{ (by canceling the } x\text{'s.)}$$

$$= \frac{-xy + x}{y}$$

Don't forget, you can't cancel the y's here! ■

Example 8: Simplify the expression $\dfrac{\dfrac{x^2 y}{z}}{\dfrac{x y^2}{z^3}}$.

Solution: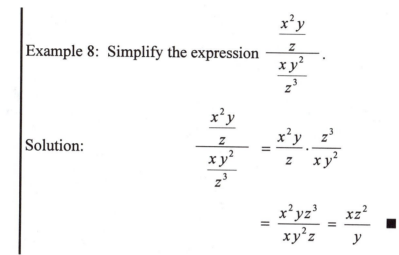

$$\frac{\dfrac{x^2 y}{z}}{\dfrac{x y^2}{z^3}} = \frac{x^2 y}{z} \cdot \frac{z^3}{x y^2}$$

$$= \frac{x^2 y z^3}{x y^2 z} = \frac{x z^2}{y} \quad ■$$

Again, if you are confused by the exponents, skip ahead to section 2.1 and then return here.

Exercises 1.3   In Exercises 1-6, multiply out and reduce to lowest terms. (Some answers may have more than one form.)

1) $\dfrac{2}{3} \cdot \dfrac{3}{4}$

2) $\dfrac{5}{16} \cdot \dfrac{8}{10}$

3) $\dfrac{-1}{3} \cdot \dfrac{-9}{5}$

4) $\dfrac{\dfrac{4}{75}}{\dfrac{8}{25}}$

5) $\dfrac{\dfrac{-7}{51}}{\dfrac{3}{12}}$

6) $\dfrac{\dfrac{3\pi}{7}}{\dfrac{2\pi}{3}}$

7)   Use your calculator to find $\dfrac{3.172}{4.8} \cdot \dfrac{2}{1.37}$, rounded to three decimal places.

8)   Use your calculator to find $\dfrac{\dfrac{5.8}{7.21}}{\dfrac{-1.23}{8.7}}$, rounded to two decimal places.

Simplify the following:

9) $\dfrac{7x}{3y} \cdot \dfrac{3y+2}{x}$

10) $\left(\dfrac{x+2}{1+y}\right) \cdot \left(\dfrac{1-y}{x}\right)$

11) $\dfrac{\dfrac{xy}{w}}{\dfrac{xy-2x}{w}}$

12) $\dfrac{xy}{wz} \cdot \dfrac{w^2 z}{x^2 y^2}$

13) $\dfrac{\dfrac{xy}{(x+y)}}{\dfrac{x^2 y}{(x+y)^3}}$

14) $\dfrac{\dfrac{xy}{(x-y)}}{\dfrac{x^2}{y} \cdot \dfrac{y^3}{x}}$

1.4 Adding and Subtracting Fractions

Adding and subtracting fractions is easy when the denominators are all the same. For example,

$$\frac{2}{5} + \frac{17}{5} - \frac{4}{5} = \frac{2 + 17 - 4}{5} = \frac{15}{5} = 3.$$

But what if the fractions don't all have the same denominator? Then you first need to rewrite the fractions so that they all have the same denominator, called a <u>common denominator</u>.

For example, if you want to add $\frac{2}{3} + \frac{4}{5}$, you can use a common denominator of 15. So:

$$\frac{2}{3} + \frac{4}{5} = \frac{10}{15} + \frac{12}{15} = \frac{22}{15}.$$

Remember, to get a common denominator you can always use the product of the individual denominators ($3 \cdot 5 = 15$), but sometimes even a smaller number will do it. Then, to get the new numerator,

$$\frac{2}{3} = \frac{?}{15},$$

divide 3 into 15, to get 5, then multiply by 2, to get 10. Try it: $\frac{3}{7} = \frac{?}{42}$. (We're doing the opposite of canceling common factors. We are in effect multiplying both top and bottom by 5, which means we are not changing the value of the fraction, only its form.)

Example 1:   a)   Simplify $\dfrac{3}{5} + \dfrac{1}{2} - \dfrac{2}{3}$.

Solution: $5 \cdot 2 \cdot 3 = 30$ will serve as a common denominator. So:

$$\dfrac{3}{5} + \dfrac{1}{2} - \dfrac{2}{3} = \dfrac{18}{30} + \dfrac{15}{30} - \dfrac{20}{30}$$

$$= \dfrac{18 + 15 - 20}{30} = \dfrac{13}{30}$$

b)   Simplify $\dfrac{1}{6} - \dfrac{1}{9}$.

Solution: We could use $6 \cdot 9 = 54$ as a common denominator, but even 18 will do nicely, because <u>both 6 and 9 divide evenly into 18</u>. So:

$$\dfrac{1}{6} - \dfrac{1}{9} = \dfrac{3}{18} - \dfrac{2}{18} = \dfrac{1}{18}$$

c)   Simplify $\dfrac{\dfrac{1}{2} + \dfrac{3}{4}}{\dfrac{1}{3} - \dfrac{1}{6}}$.

Solution: $\dfrac{\dfrac{1}{2} + \dfrac{3}{4}}{\dfrac{1}{3} - \dfrac{1}{6}} = \dfrac{\dfrac{2 + 3}{4}}{\dfrac{2 - 1}{6}} = \dfrac{\dfrac{5}{4}}{\dfrac{1}{6}} = \dfrac{5}{4} \cdot \dfrac{6}{1} = \dfrac{15}{2}$ ∎

Sometimes you'll want to use your calculator to add fractions because it makes life much simpler.

Example 2:   Use your calculator to find $\dfrac{1.8}{3.172 - 2.019}$, rounded to two decimal places.

Solution: Now you will have to use the parenthesis keys in order to calculate correctly. Pressing

1.8 $\boxed{\div}$ $\boxed{(}$ 3.172 $\boxed{-}$ 2.019 $\boxed{)}$ $\boxed{\text{ENTER}}$

you will find out that

$$\dfrac{1.8}{3.172 - 2.019} = 1.56114484 .$$

So 1.56 is the correct answer.

If you have not used the parenthesis keys in this case, the order of operations would have caused a wrong calculation of -1.45 (Try it!). ∎

Knowing when to use the parenthesis keys on your calculator requires your understanding exactly the order of calculations your calculator uses. Again, when in doubt use your parenthesis keys!

You may also have to be comfortable with adding and subtracting fractional expressions involving variables.

Example 3:   Simplify $\dfrac{xy}{z} + \dfrac{x}{5}$.

Solution: Using a common denominator of $5z$ we write

$$\dfrac{xy}{z} = \dfrac{5xy}{5z} \quad \text{and} \quad \dfrac{x}{5} = \dfrac{xz}{5z}$$

and so,

$$\frac{xy}{z} + \frac{x}{5} = \frac{5xy}{5z} + \frac{xz}{5z} = \frac{5xy + xz}{5z}.$$

Remember you can't cancel the 5's! ■

Example 4:

$$\frac{z}{y^2} - \frac{x}{z^3 y} = \frac{z^4}{y^2 z^3} - \frac{xy}{y^2 z^3} = \frac{z^4 - xy}{y^2 z^3} \quad ■$$

Example 5:

$$\frac{\dfrac{3a + 2b}{5ab}}{\dfrac{a}{b} - \dfrac{b}{a}} = \frac{\dfrac{3a + 2b}{5ab}}{\dfrac{a^2 - b^2}{ab}}$$

$$= \frac{3a + 2b}{5ab} \cdot \frac{ab}{a^2 - b^2}$$

Cancel ab, then multiply out the bottom to get

$$\frac{3a + 2b}{5a^2 - 5b^2}. \quad ■$$

Exercises 1.4 Express as a single fraction and simplify:

1) $\dfrac{1}{5} + \dfrac{3}{5}$

2) $\dfrac{1}{2} + \dfrac{1}{3}$

3) $\dfrac{1}{3} - \dfrac{1}{6}$

4) $\dfrac{1}{3} + \dfrac{1}{4}$

5) $\dfrac{7}{6} + \dfrac{5}{24}$

6) $\dfrac{2}{5} - \dfrac{1}{2} + \dfrac{1}{3}$

7) $\dfrac{1}{2} - \dfrac{1}{4} + \dfrac{1}{8} - \dfrac{1}{16}$

8) $\dfrac{1}{2} + \dfrac{4}{3} - \dfrac{2}{5} - \dfrac{3}{15}$

9) $\dfrac{5}{6} - \dfrac{4}{3} + \dfrac{2}{9} - \dfrac{3}{2}$

10) $\dfrac{\frac{1}{3} + \frac{2}{5}}{\frac{3}{2}}$

11) $\dfrac{\frac{1}{4} - \frac{2}{3}}{\frac{3}{2} - \frac{2}{5}}$

12) $\dfrac{\frac{2}{7} + \frac{1}{3}}{\frac{4}{3} + \frac{2}{5}} + \dfrac{1}{3}$

13) Use your calculator to find $\dfrac{1.9}{0.0137 - 0.00249}$, rounded to two decimal places.

14) Use your calculator to find $\dfrac{3.898}{0.14} - \dfrac{1.745}{2.5356}$, rounded to four decimal places.

15) Use your calculator to find $\dfrac{1.87}{3.33} + \dfrac{2.7}{2} - 1.32(2.45 - 1.338)$, rounded to two decimal places.

16) Use your calculator to find $\dfrac{3.87}{-1.2} + \dfrac{.007}{5.2 - 5.103}$, rounded to three decimal places.

Simplify the following:

17) $\dfrac{1}{x} + \dfrac{1}{y}$

18) $\dfrac{1}{y} - \dfrac{1}{x}$

19) $\dfrac{4}{x} - \dfrac{2}{y} + \dfrac{1}{z}$

20) $\dfrac{1}{x} - \dfrac{x+1}{xy} + \dfrac{x-2}{xz}$

21) $\dfrac{\frac{1}{y} - \frac{x}{z}}{\frac{1}{z} - \frac{1}{x}}$

22) $\dfrac{\frac{1}{st} - \frac{1}{w}}{\frac{1}{tw} - \frac{2}{s}}$

1.5 <u>Percent</u>

Percent, represented by the symbol %, means "per hundred." So 5% of 400 means

5 one-hundredths of 400, which is $\left(\dfrac{5}{100}\right)(400) = 20$. Always, x % of y is $\left(\dfrac{x}{100}\right)(y)$.

Example 1:   Find 15% of 90.

Solution: $\dfrac{15}{100} \cdot 90 = \dfrac{15 \cdot 9}{10} = \dfrac{135}{10} = 13.5$ ∎

Example 2:   Find 1% of 320.

Solution: $\dfrac{1}{100} \cdot 320 = \dfrac{320}{100} = 3.2$

Note that 1% of any number is always $\dfrac{1}{100} = .01$ of the number, so move the decimal point over two places to the left! (For example, 1% of 4567 is 45.67 and 1% of 378.2 is 3.782 .) ∎

Example 3:   Find $\frac{1}{3}$% of 930.

Solution: Since 1% of 930 is 9.3, $\frac{1}{3}$% of 930 would be $\frac{1}{3}$ of 9.3, or 3.1. Alternatively:

$$\dfrac{\frac{1}{3}}{100} \cdot 930 = \dfrac{1}{3} \cdot \dfrac{1}{100} \cdot \dfrac{930}{1} = \dfrac{930}{300} = \dfrac{93}{30} = \dfrac{31}{10} = 3.1$$ ∎

Example 4: Find 250% of 150.

Solution: $\dfrac{250}{100}\cdot 150 = \dfrac{250\cdot 15}{10} = 25\cdot 15 = 375$ ∎

Percentages are often used to describe ratios pertaining to results in survey questionnaires. For example, if 8 of 20 people prefer creamy peanut butter to crunchy, what percent is that? Well, 8 of 20 is the same ratio as 40 of 100, so 8 is 40% of 20. Notice that we can calculate like this:

$$\frac{8}{20} = \left(\frac{8}{20}\right)(100)\% = \frac{800}{20}\% = 40\%$$

Example 5: Suppose that out of 523 people, 388 responded "Yes" to the question "Are you grouchy in the morning?" What percentage of people in this survey admit that they are grouchy in the morning?

Solution: The admitted morning grouches make up a fraction of $\dfrac{388}{523}$ of the group, which equals

$$\frac{388}{523}\cdot 100\% = (0.74187)100\%$$
$$= 74.187\%.$$

So the answer is approximately 74%. ∎

Example 6:   Ed has a blood alcohol level of 0.15%. If his total blood volume is approximately five liters, then how much pure alcohol is in his blood stream?

Solution: We wish to know 0.15% of five liters. Don't let the decimal throw you.   (If we wanted 43% of 5, we'd calculate $\left(\frac{43}{100}\right)(5)$, right?) So here we have

$$\left(\frac{0.15}{100}\right)(5) = \frac{0.75}{100} = 0.0075,$$

in liters of course. ■

Example 7:   A CD usually sells for $15.99. Shower Records is having a big sale, offering 60% off. How many CD's can you get for $20, and exactly how much would you pay (assuming no sales tax)?

Solution: A 60%-off sale means that the CD's will cost 40% of their original price. Now 40% of $15.99 is

$$\frac{40}{100} \cdot (\$15.99) = \frac{4 \cdot (\$15.99)}{10} = \$6.396 = \$6.40$$

(rounded up to the nearest cent). So you can buy three CD's, and $19.20 is what you'll pay. ■

Example 8:   The new federal budget has a debt of 147 billion dollars, which is 30% below last year's deficit. What was last year's deficit?

Solution: This means that $147 billion is 70% of last year's deficit. (Right?) Let's put x = last year's deficit.

So 70% of x is 147 billion, that is

$$\frac{70}{100} \cdot x = 147 \text{ billion.}$$

Solving for x: $x = \frac{100}{70} \cdot (147 \text{ billion})$

$$= 210 \text{ billion}$$

So last year's deficit was 210 billion dollars. ∎

Exercises 1.5

1) Find 25% of 200.

2) What is 3.3% of 7.1?

3) The number 587 is 45% of what number?

4) Suppose 8500 people make up a survey sample for a questionnaire on taxes. If 4487 of the people are female and the remaining 4013 are male, then what percent of the sample are female? What percent are male?

5) When asked "Do you study less than six hours a week?", 1605 of 2540 students responded yes, all the rest responded no. What percent of the students say that they study six or more hours a week?

6) The final grades for Dr. Jekyll's Stat class are given in the table below.

Grade	Number
A	2
B	5
C	12
D	27
F	33

a) What percentage of his class received A's for final grades? b) B's ? c) C's ?
d) D's ? e) F's ?   Round your answers to one decimal place.

7) The bookstore is having an inventory sale with a 25% reduction in prices. a) How much will a $16.98 sweatshirt cost? b) How about a $1499.00 computer?

8) Joe heard that the government was selling foreclosed land at 35% off. He went to look at a farm going for the reduced price of $97,500. Joe bought the farm. How much did he save?

9) Diana borrowed $100,000 to start a new business. The loan has an interest rate of "prime plus two," with only the interest to be paid once a year. (Nice if you can get it.) The prime rate in the first year was 6 %, and so her interest rate was 8 %.

a) How much did she pay at the end of the first year?

b) The prime rate in the second year rose to 9 %. How much did she pay after the second year?

c) How much did the interest payment increase from the first year to the second?

d) What was the percentage increase in the monthly payment?

1.6 Intervals and Inequalities

The real numbers are associated with points on a line called the number line, as shown below:

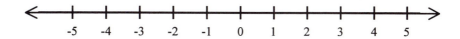

The integers are shown to help you locate the other numbers.   All real numbers correspond to exactly one point on this line. When thinking of real numbers, we can think of them equally well as points on this line.

Definition: Let a and b be real numbers. We write:

i) $a < b$ if a is less than b. (This means that a is to the left of b on the number line.)

ii) $a \leq b$ if $a < b$ or $a = b$.

iii) $a > b$ if a is greater than b. (This means that a is to the right of b on the number line.)

iv) $a \geq b$ if $a > b$ or $a = b$.

An inequality is a statement involving any of these signs.

Example 1: All of the following are true. Do you agree?

a) $3 < 5$

b) $3 \leq 5$ (because $3 < 5$)

c) $5 \leq 5$ (because $5 = 5$)

d) $-5 < -3$

e) $x^2 \geq 0$ for all real numbers x. ∎

An interval is just a connected piece of the number line. For example, the set of all real numbers between 0 and 1, including 0 and 1, is called a closed interval and is denoted [0,1].   If you don't mean to include 0 and 1, you have an open interval and express it by using round parentheses, (0,1).   If 0 is to be included, but not 1, you write [0,1). Each of these intervals can also be shown on the number line, or expressed as a pair of inequalities. In general, if $a < b$:

Interval Notation	**Number Line**	**Inequalities**
$[a,b]$		$a \leq x \leq b$
(a,b)		$a < x < b$
$[a,b)$		$a \leq x < b$
$(a,b]$		$a < x \leq b$

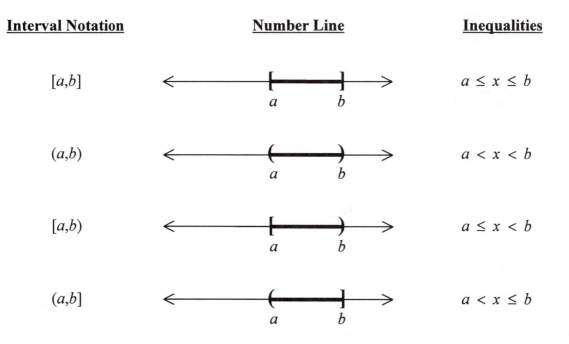

In all of these examples, the number a is called the left endpoint, and b is called the right endpoint. In your statistics course you will meet so-called confidence intervals, which are of the form $a < x < b$, and illustrated above (second from the top). Your calculator will display these intervals as (a,b).

Example 2:   Express the interval $(-1.2, 2.5)$ using inequalities and graph the interval on the number line.

Solution: Here, the left endpoint is -1.2 and the right endpoint is 2.5 so the inequality is $-1.2 < x < 2.5$. Note that since the interval has round parentheses on each end, the inequalities are simple "less than" signs. The interval looks like:

since -1.2 is a little bit to the left of -1, and 2.5 is halfway between 2 and 3. ■

All of the above are called <u>finite intervals</u> because their length is finite. There are also <u>infinite intervals</u>. Infinite just means "no end," that is, the interval keeps on going, without end. This is expressed by the infinity symbol ∞ on the right and $-\infty$ on the left. Here are some examples.

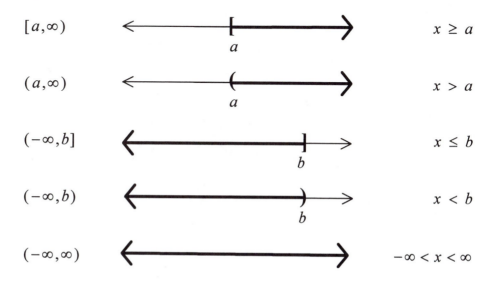

$[a,\infty)$		$x \geq a$
(a,∞)		$x > a$
$(-\infty,b]$		$x \leq b$
$(-\infty,b)$		$x < b$
$(-\infty,\infty)$		$-\infty < x < \infty$

In case you ever need the following material on union and intersection, we have added it here.

Definition:　Let A and B be two sets of objects of any sort.

a)　The set of all objects that are in **both** A and B is called "$\underline{A\ intersection\ B}$," and is denoted $A \cap B$.

b)　The set of all objects that are in **either** A **or** B (**or both**) is called "$\underline{A\ union\ B}$," and is denoted $A \cup B$.

The above definitions can be applied to intervals, and they often are. Consider the following:

Example 3:　a)　$(-2,1) \cap (0,2) = (0,1)$

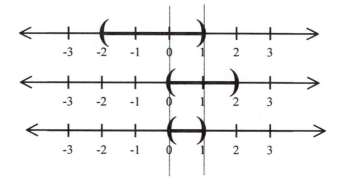

b)　$(-\infty,5) \cap (3,\infty)$ is the set of all numbers that are less than 5 and at the same time greater than 3. So it is the set $(3,5)$.

c) $(5,10) \cap (21,\infty) = \varnothing$, the empty set, since there
are no numbers that are both less than 10 and at the
same time greater than 21.

d) $(-\infty,5) \cup (4,\infty) = (-\infty,\infty)$, the complete
number line.

e) $(-\infty,5] \cup (10,21)$ can't be written in a more
compact form. ■

In the last example we see the use of union to describe sets of points on the
number line which are not connected. We often speak of a set of numbers consisting of
two or more intervals, finite or infinite. This is when the union sign comes in handy!

Exercises 1.6

1) Draw the number line, and locate the following numbers:

a) $\dfrac{-126}{106}$ b) $-3 + \dfrac{10}{27}$ c) $\dfrac{3.14}{2} - 1$

2) Represent the following sets of numbers using interval and number line notation.

a) $-1 \le x \le 3$ b) $-1 < x \le 3$ c) $-3 \le x < 1$

2) continued

d)	$-3 \le x \le 4$	e)	$-\dfrac{1}{2} < x \le \sqrt{2}$	f)	$3.14 \le x \le 5$
g)	$0 < x$	h)	$3 \le x$	i)	$x < -4$
j)	$-0.091 \le x$	k)	$x < 5$	l)	$x \le 3$

3) Represent the following intervals using inequalities.

a)	$(3,7)$	b)	$(-4,-1]$	c)	$(-\infty,19]$
d)	$[2,10)$	e)	$[-2,-1]$	f)	$(-2.3,5)$
g)	$(-0.99,0.99)$	h)	$(3.1,18.75)$		

4) Simplify if possible:

a)	$(-2,2) \cup (1,5)$	b)	$(-10,-1) \cup (0,5)$
c)	$(-\infty,5) \cap [3,\infty)$	d)	$(-\infty,5) \cup [3,\infty)$
e)	$(-\infty,-2) \cap [-2,\infty)$	f)	$(-\infty,\infty) \cap [4,7]$

Chapter 2

Exponents

2.1 Exponents

Positive whole number exponents are a simple mathematical notation to represent repeated multiplication by the same factor. But you can also have negative numbers and 0 as exponents. Consider the table below.

	3^4	$= 3 \cdot 3 \cdot 3 \cdot 3$	$=81$
	3^3	$= 3 \cdot 3 \cdot 3$	$=27$
Exponents increase by 1.	3^2	$= 3 \cdot 3$	$=9$
	3^1	$= 3$	$=3$
↑	3^0	$= 1$	$=1$
Numbers are multiplied by 3.	3^{-1}	$= \frac{1}{3}$	$= \frac{1}{3}$
	3^{-2}	$= \frac{1}{3^2}$	$= \frac{1}{9}$
	3^{-3}	$= \frac{1}{3^3}$	$= \frac{1}{27}$
	3^{-4}	$= \frac{1}{3^4}$	$= \frac{1}{81}$

Exponents increase by 1.

↑

Numbers are multiplied by 3.

Exponents decrease by 1.

↓

Numbers are divided by 3.

See the pattern? Let's summarize: $3^n = \underbrace{3 \cdot 3 \cdot 3 \cdot \cdots \cdot 3 \cdot 3}_{n \text{ factors}}$, $3^{-n} = \dfrac{1}{3^n}$, and $3^0 = 1$. In general, given any number x and any positive integer n,

$$x^n = \underbrace{x \cdot x \cdot x \cdot \cdots \cdot x \cdot x}_{n \text{ factors}} ,$$

$$x^{-n} = \frac{1}{x^n}, \text{ for } x \neq 0,$$

and

$$x^0 = 1, \text{ for } x \neq 0.$$

In all of these rules, the number x is called the <u>base</u>, while the number n is called the <u>exponent</u>.

<u>Properties of Exponents:</u> Let a and b be any real numbers. Then, in each of the following, if the expressions on both sides exist, they will be equal.

1. $a^r \cdot a^s = a^{r+s}$ 2. $\dfrac{a^r}{a^s} = a^{r-s}$

3. $(a^r)^s = a^{r \cdot s}$ 4. $(ab)^r = a^r b^r$

5. $\left(\dfrac{a}{b}\right)^r = \dfrac{a^r}{b^r}$ 6. $\left(\dfrac{a}{b}\right)^{-r} = \left(\dfrac{b}{a}\right)^r$

Example 1:   a) $\dfrac{4^2 - 1}{3^3 - 2^2} = \dfrac{16 - 1}{27 - 4} = \dfrac{15}{23}$

b) $5^{-1} + 3^{-1} = \dfrac{1}{5} + \dfrac{1}{3} = \dfrac{3 + 5}{15} = \dfrac{8}{15}$

c) $\dfrac{3 \cdot 8^2}{9 \cdot 8^3} = \dfrac{3}{9} \cdot \dfrac{8^2}{8^3} = \dfrac{1}{3} \cdot \dfrac{1}{8} = \dfrac{1}{24}$ ■

Example 2: $\left(\dfrac{x^{-2}}{x^8}\right)^{-2} = \left(x^{-10}\right)^{-2}$ by property #2

$= x^{20}$ by property #3   ■

Example 3: $\dfrac{1}{a^3} - \left(\dfrac{1}{a^5} - \dfrac{1}{a^2}\right) = \dfrac{1}{a^3} - \dfrac{1}{a^5} + \dfrac{1}{a^2}$

$= \dfrac{a^2 - 1 + a^3}{a^5}$ ■

Example 4: $\dfrac{x^2 y^5}{x^{-3}} \div \dfrac{x^{-5} y^4}{x^3} = \left(\dfrac{x^2}{x^{-3}} y^5\right) \div \left(\dfrac{x^{-5}}{x^3} y^4\right)$

(It's useful to separate the x's and the y's.)

$= x^5 y^5 \div x^{-8} y^4$ by property #2

$= \dfrac{x^5 y^5}{x^{-8} y^4} = x^{13} y$ ■

The TI-83 has an exponent key that looks like $\boxed{\wedge}$. Scientific calculators have a key that looks like $\boxed{y^x}$.   Some brands, like Casio, have an $\boxed{x^y}$ key. If you're calculating powers, these keys are lifesavers.

Example 5:   a) Use you calculator to compute $(1.07)^{10}$, rounded to three decimal places. b) If you invest $1000 at 7% interest, compounded annually, what is the value of your investment in ten years? (Round your answer to the nearest penny.)

Solution: a) For the TI-83 press:

$$1.07 \quad \boxed{\wedge} \quad 10 \quad \boxed{\text{ENTER}}$$

For other calculators press:

$$1.07 \quad \boxed{y^x} \quad 10 \quad \boxed{=}$$

The result on your calculator should be that

$$(1.07)^{10} = 1.967151357 ,$$

which becomes 1.967 when rounded.

b) Well, after one year the value would be

$$
\begin{aligned}
1000 + 7\% \text{ of } 1000 &= 1000 + (0.07)(1000) \\
&= 1000(1 + 0.07) \\
&= 1000(1.07),
\end{aligned}
$$

which is just 1.07 times the original amount. So after two years, it would be 1.07 times the amount after one year or

$$[1000(1.07)](1.07) = 1000(1.07)^2 .$$

After three years, it would be

$$[1000(1.07)^2](1.07) = 1000(1.07)^3.$$

See the pattern? [Each year an extra factor of 1.07 is brought in.] Hence, after ten years the value of the investment would be

$$1000(1.07)^{10} = \$1967,$$

where we have used the result in part (a). ■

Exercises 2.1 Simplify the following expressions:

1) $\dfrac{4^{-1}5^2}{2^3 3^{-2}}$

2) $\dfrac{3^5 2^3}{4^2 3^3}$

3) $\dfrac{5^3}{3^{-1}5^2 + 4^{-1}5^3}$

4) $4^3 \left(\dfrac{1}{4}\right)^2 3^{-4}$

5) $\dfrac{1}{2^{-3}} - \dfrac{1}{2} + \dfrac{1}{5^{-2}}$

6) $x^2 y^{-2} z^3 x^{-2} y^3 z^5$

7) $\left(2^2\right)^{-1}$

8) $\left(x^2\right)^{34}$

9) $\left(\dfrac{1}{x^2}\right)^{34}$

10) Use your calculator to compute $(1.134)^3$, rounded to five decimal places.

11) Use your calculator to compute $(-2.034)^4$, rounded to three decimal places.

12) Use your calculator to compute $(-0.934)^5$, rounded to three decimal places.

Simplify:

13) $\dfrac{x^{-1}y^2}{y^2 x^{-2}}$

14) $\dfrac{\left(x^2 y^{-3}\right)^2}{\left(y^{-3} x^{-2}\right)^{-2}}$

15) $\dfrac{x^2 y}{x^3} \div \dfrac{x^{-3} y^6}{y^4}$

16) $\left(x^{-1} + y^{-1}\right)^{-1}$

2.2 Scientific Notation

Decimal notation is OK for many numbers, but for really large or small numbers it is a big nuisance. Take the deficit (please!): 210 billion = 210,000,000,000 (in decimal form). But notice that all the zeros really represent multiplication by 10 so that

$$210{,}000{,}000{,}000 \;=\; 21 \times 10^{10} \;=\; 2.1 \times 10^{11}.$$

This last expression using the factors of 10 is called <u>scientific notation</u>. Notice that scientific notation calls for:

$$\pm \begin{bmatrix} \text{a number greater than or equal} \\ \text{to 1, but less than 10, written} \\ \text{in decimal form} \end{bmatrix} \times 10^{\left(\begin{array}{l} \text{some integer, positive} \\ \text{or negative, or 0} \end{array} \right)}$$

For example, you would write .0000536 in scientific notation as 5.36×10^{-5}. You would write 6437.8 as 6.4378×10^{3}. Just count the number of positions you have to move the decimal point so that the resulting number is between 1 and 10. If you have to move the decimal point left n places, the exponent is n; if you move it n places to the right, the exponent is $-n$.

Remarks: 1) If the decimal component of a number written in scientific notation has n digits, the number is said to have n <u>significant digits</u>. So the number 4.5601×10^{3} has five significant digits.

2) On most calculators, very large or small numbers are given in something like scientific notation. For example, the number 3.28×10^{-9} may look something like 3.2800000E⁻09, or $3.28^{\ -09}$.

Example 1: The number of molecules in a mole (a mole??) of gas is called Avogadro's number. It is equal to 6.023×10^{23} Write it in decimal [nonscientific] form.

Solution: 6.023×10^{23} = $602,300,000,000,000,000,000,000$

See why scientific notation is so handy! ∎

The above material may be all you need to know about scientific notation, but if you need to calculate with it, read on!

Example 2: The speed of light (in a vacuum) is 186,000 miles per second. How big is a light-year (the distance light travels in one year)? Express the answer in scientific notation, rounded to three digits.

Solution: First, the speed of light written in scientific notation is 1.86×10^5. Now there are 365 days a year, so there are

365 days × 24 (hr/day) = 8760 hours per year, and

365 (days/year) × 24 (hr/day) × 3600 (sec/hr) = 31,536,000 seconds per year. In scientific notation, this number is 3.1536×10^7 seconds.

As you know, in each of these seconds, the light travels 186,000 miles. So in one year, it travels (31,536,000)(186,000) miles, and hence

1 light-year = $(3.1536 \times 10^7)(1.86 \times 10^5)$

$$= (3.1536)(1.86) \times (10^7)(10^5)$$

$$= 5.865696 \times 10^{12}$$

$$\cong 5.87 \times 10^{12} \text{ miles. } \blacksquare$$

Remark :   This is in scientific notation since the first factor (5.87) is between 1 and 10. If it had not been between 1 and 10, you would have to write it differently. For example, if the product had ended up being 22.87×10^{12}, you would have to write it as 2.287×10^{13}.

Example 3:   Calculate the product $(3.01 \times 10^{-3}) \cdot (4 \times 10^5)$ with and without a calculator, and express it in scientific notation.

Solution: a) Without a calculator we write

$$
\begin{aligned}
(3.01 \times 10^{-3}) \cdot (4 \times 10^5) &= 3.01 \times 10^{-3} \times 4 \times 10^5 \\
&= (3.01 \times 4) \times (10^{-3} \times 10^5) \\
&= 12.04 \times 10^2
\end{aligned}
$$

Note that in this case we have to go the extra step mentioned in the Remark above. That is, the coefficient of the power of 10 is actually greater than 10, so we move the decimal over one more place and adjust the power of 10 to get 1.204×10^3.

b) To compute with numbers written in scientific notation on a calculator, one could use the exponentiation key $\boxed{\wedge}$ (or $\boxed{y^x}$ or $\boxed{x^y}$) as in the last section, but because scientific notation is so common there is a special key that is used to enter the exponent.

To express multiplication by 10^{-3}:

[2nd] [,] [(−)] 3   (on the TI-83)

[EE] [+/−] 3   (on most scientific calculators)

Note that the [(−)] (or the [+/−]) key is used to make the exponent negative: −3 .

Here we go. For the TI-83 you press:

3.01 [2nd] [,] [(−)] 3 [×] 4 [2nd] [,] 5 [ENTER]

For most other calculators press:

3.01 [EE] [+/−] 3 [×] 4 [EE] 5 [=]

In either case we get the desired solution 1.204E3 .

Note: If your calculator is set in the normal mode, then the answer 1204 might appear. ∎

Editorial Comment: Sometimes using your head is easier and faster than using a calculator. In fact if you use a calculator, you should always do at least a rough calculation in your head, to make sure the answer makes sense. This can avoid those embarrassing bloopers.

Exercises 2.2

1)   Express the following numbers in scientific notation, rounded to three digits.

   a)   382935.9938   b) -0.000724

   c)   3.000001   d)   200.001

2)   Convert the following numbers from scientific notation to decimal notation.

   a) 3.4178×10^3 b) -1.0915×10^2

   c) 5.4378×10^{-6} d) -4.76×10^{-2}

   e) 8.00001×10^{10}

3)   Compute the following and express in scientific notation, rounded to three digits. You may use your calculator.

   a) $(2.35 \times 10^5) \times (4.032 \times 10^2)$ b) $(-6.15 \times 10^{-2}) \times (5.032 \times 10^6)$

   c) $(-5.001 \times 10^{-2}) \times (-7.001 \times 10^{-99})$ d) $\dfrac{3.24 \times 10^2}{4.23 \times 10^3}$

   e) $\dfrac{-1.33 \times 10^{-2}}{7.9 \times 10^5}$ f) $(3.82 \times 10^{-1})^3$

4) a) You invest $ 2500 in a fund that earns 6% annually. What will its value be in 20 years?

   b) If the cost-of-living increased at a constant rate for these 20 years at 4% annually, then the "real" value of your investment, meaning your money's purchasing power, increases at 2% annually.   What is the increase in purchasing power in 20 years?   What is the percentage increase?

2.3 Square Roots

We know that 3 is a square root of 9 since $3^2 = 9$. We also know that -3 is a square root of 9 since $(-3)^2 = 9$ too. So the number 9 has two square roots. In fact, any positive number a has two square roots, a positive one and a negative one. The positive one is denoted $\sqrt{a}$, and the negative one is denoted $-\sqrt{a}$. So 25 has two square roots, $\sqrt{25}$ and $-\sqrt{25}$, also known as 5 and -5. (Sometimes $\sqrt{a}$ is written as $\sqrt[2]{a}$.)

Does -16 have a square root? Well, if we square any real number, whether positive or negative or zero, we'll never get -16. So -16 has no real number as a square root. We express this in shorthand by saying "-16 has no real square root." In fact, no negative number has a real square root.

The following properties of roots come from the properties of exponents, and are often very useful in calculations. Note not only the properties which can be used, but also the pitfalls, which must be avoided.

Properties of Roots: Let $a, b > 0$. Then:

$$1. \qquad \sqrt{ab} \;=\; \sqrt{a}\,\sqrt{b}$$

$$2. \qquad \sqrt{\frac{a}{b}} \;=\; \frac{\sqrt{a}}{\sqrt{b}}$$

But, notice that:

3. $\sqrt{a+b}$ is NOT equal to $\sqrt{a}+\sqrt{b}$ (at least not in general).

4. $\sqrt{a-b}$ is NOT equal to $\sqrt{a}-\sqrt{b}$ (at least not in general).

Example 1:   Simplify: a) $\sqrt{9\cdot16}$ b) $\sqrt{\dfrac{100}{4}}$

Solution: a) $\sqrt{9\cdot16} = \sqrt{9}\cdot\sqrt{16} = 3\cdot4 = 12$

Notice that this could have been done differently, avoiding the properties of roots, taking somewhat more effort:

$$\sqrt{9\cdot16} = \sqrt{144} = 12$$

b) $\sqrt{\dfrac{100}{4}} = \dfrac{\sqrt{100}}{\sqrt{4}} = \dfrac{10}{2} = 5 \quad\blacksquare$

Example 2:   Using your calculator, approximate $\sqrt{27}$ to three decimal places.

Solution: The square root key on the TI-83 is $\boxed{\text{2nd}}\ \boxed{x^2}$. When pressing that you will see $\sqrt{\ (}$ on the screen. To calculate $\sqrt{27}$ you would press

| 2nd | | x^2 | 27 | ENTER |

and see the answer 5.196152423, which is 5.196 when rounded to three decimal places.

To use the square root key on other calculators you simply enter the number then press the $\boxed{\sqrt{}}$ key. ■

Example 3:   Simplify:

a) $\sqrt{9 + 16}$ b) $\sqrt{-16 + 49}$ c) $\sqrt{1^2 + 2^2 + 3^2 + 4^2}$

Solution: a) $\sqrt{9 + 16} = \sqrt{25} = 5$

Notice that it's NOT EQUAL to $\sqrt{9} + \sqrt{16}$, because that is $3 + 4$ which is 7, not 5. Get the picture?

b) $\sqrt{-16 + 49} = \sqrt{33} = 5.7$,   rounded to one decimal place.

c) $\sqrt{1^2 + 2^2 + 3^2 + 4^2}$ is NOT equal to $1 + 2 + 3 + 4$; rather it equals $\sqrt{1 + 4 + 9 + 16} = \sqrt{30} = 5.5$ rounded to one decimal place. ■

Exercises 2.3   In Exercises 1-8, simplify the following, if they exist, rounding to two decimal places where necessary.

1) $\sqrt{144}$ 2) $\sqrt{\dfrac{1}{9}}$ 3) $\sqrt{\dfrac{4}{49}}$ 4) $\sqrt{5 \cdot 3}$

5) $\sqrt{5^2 - 36}$ 6) $\sqrt{\dfrac{5}{\sqrt{7.8}}}$ 7) $\sqrt{4(1.3 + 2.8)}$ 8) $\sqrt{5^2 + 12^2}$

9)   Suppose $a = 1.2$, $b = 0.1$, $c = -1.3$, and $d = 0.4$, then compute

$$\sqrt{a^2 + b^2 + c^2 + d^2} \ .$$

10) a) If $a > 0$ and $b > 0$, is $\sqrt{a^2 + b^2}$ always the same as $a + b$? Justify your answer.

b) If $a > 0$ and $b > 0$, is $\sqrt{a^2 + b^2}$ ever the same as $a + b$? Justify your answer.

11) If $x^2 + y^2 = 25$, can we conclude that $x + y = 5$? Why or why not?

Chapter 3

Subscripts and Summation

3.1 Mean, Subscripts, and $\sum$-Notation

If your test results in a course are 87, 49, 67, and 93, it's easy to find the mean. Add them up, and divide by 4, to get 74. (This is also called the average.) In statistics it's customary to use only the word mean. We want to generalize this notion of mean. First of all, we want to consider not just four quantities, but any number of them. Let's call the number of quantities n. We'll call the first quantity x_1, the second one x_2, the third one x_3, etc., and the last one x_n. These little numbers written below are called subscripts or indices (singular: index). The reason for using subscripts is the ease with which one can refer to any particular quantity: x_{12} is the 12th quantity of interest. If instead we used the alphabet a, b, c, d, etc., for variable names, we couldn't easily write down any general expressions like the following. Also we'd be stuck anytime there are more than 26 of them.

Definition: Let x_1, x_2, x_3, $\cdots$, x_n be real numbers. The mean of this set of numbers is defined to be

$$\frac{x_1 + x_2 + x_3 + \cdots + x_n}{n}.$$

The expression $x_1 + x_2 + x_3 + \cdots + x_n$ in this definition denotes the sum of the n numbers. It's rather long and a bit awkward. Since this expression pops up so often in calculations, another notation is also used. The symbol Σ (the capital Greek letter sigma, standing for S) is used to indicate summation.

The symbol $\sum_{i=1}^{n} x_i$ indicates the sum of the various x_i's are to be added up, starting with $i = 1$ (namely x_1), all the way up to $i = n$ (namely x_n). That is:

$$\sum_{i=1}^{n} x_i = x_1 + x_2 + x_3 + \cdots + x_n.$$

Example 1:   Let $x_1 = 3$, $x_2 = 5$, $x_3 = -1$, and $x_4 = 17$.   Calculate $\sum_{i=1}^{4} x_i$.

Solution:

$$\sum_{i=1}^{4} x_i = x_1 + x_2 + x_3 + x_4$$

$$= 3 + 5 + (-1) + 17 = 24 \quad \blacksquare$$

In statistics, it's usually clear where the summation is to begin and end, and so it's convenient to simplify the notation to $\sum x_i$, and in fact usually to $\sum x$. Using this notation we can write

$$\text{Mean} = \frac{\sum x}{n}.$$

Example 2:   On your block, the household incomes in 1996 were: $40,000, $30,000, $60,000, $25,000, $1,030,000 (Nobel Prize money) and $18,000.   Find the mean household income.

Solution: $$\text{Mean} = \frac{\sum x}{n}$$

$$= \frac{40{,}000 + 30{,}000 + 60{,}000 + 25{,}000 + 1{,}030{,}000 + 18{,}000}{6}$$

$$= 200{,}500$$

So the mean income is $200,500. By the way, this shows that the mean income may be quite different from what we think of as "typical" income. ∎

Sometimes you will have to sum up squares of numbers, or even squares of differences of numbers. Consider the following:

Example 3:   Suppose that the student scores, x, on the first exam in your biology class are 68, 74, 59, 77, 79, 82, 99, 54, and 92.

Compute the <u>standard deviation</u> $\sigma = \sqrt{\dfrac{\sum (x-\mu)^2}{n}}$, where n is the number of test scores, and μ is the mean of the scores. (Note: This is the population standard deviation.)

Solution: Whoa, $\sqrt{\dfrac{\sum(x-\mu)^2}{n}}$ is a pretty nasty looking expression.

First we must compute the mean $\mu = \dfrac{\sum x}{n}$. We see that $n = 9$, and

$$\mu = \frac{68 + 74 + 59 + 77 + 79 + 82 + 99 + 54 + 92}{9} = \frac{684}{9}$$

$$= 76.$$

So the class average is 76. Now using this, we compute σ as follows.

First we compute $\sum(x - \mu)^2$, recalling that $\mu = 76$.

$$\sum(x - \mu)^2 = (68 - 76)^2 + (74 - 76)^2 + \cdots$$
$$= (-8)^2 + (-2)^2 + (-17)^2 + (1)^2 + (3)^2 +$$
$$+ (6)^2 + (23)^2 + (-22)^2 + (16)^2$$

Squaring out the numbers we get

$$\sum(x - \mu)^2 = 64+4+289+1+9+36+529+484+256$$
$$= 1672$$

and so $\sqrt{\dfrac{\sum(x-\mu)^2}{n}} = \sqrt{\dfrac{1672}{9}} = \sqrt{185.\overline{77}} \approx 13.63$,

when rounded to two decimal places. ■

If there is more than one variable floating around, say x and y, then it is useful to use a subscript to denote which standard deviation we are talking about. For example, σ_x represents the standard deviation of x, and σ_y represents the standard deviation of y.

If you have a TI-83, it is possible to enter raw data, or test scores in this case, and have the calculator compute the mean and standard deviation. For those students with a TI-83, read Example 4. All other students with nongraphing statistical calculators use your $\boxed{\Sigma}$ key to enter the data list. Once the list is entered, select appropriate keys to automatically compute the mean and the standard deviation. Typically these keys are "alternate functions" keys with symbols $\bar{x}$ for mean, and σ_{xn} for standard deviation.

Example 4:   Use a TI-83 to compute the mean and standard deviation for the data in Example 3.

Solution: To analyze statistical data using the TI-83, you first must enter the data into the calculator. This is done via the Stat List Editor. Press $\boxed{\text{STAT}}$ $\boxed{\text{ENTER}}$ and you will see columns labeled L1, L2, and L3 as shown below:

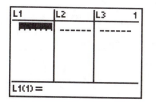

The window shows you only three columns at a time, but you can view and edit other data columns by using the left or right scroll keys. (These look like $\boxed{<}$ and $\boxed{>}$.) There are a total of 20 data columns. Six of them are named L_1, L_2, L_3, L_4, L_5, and L_6. The other 14 are for your naming pleasure, but until you are comfortable with using the Stat List Editor we suggest you use one of the six prenamed data lists. Once you have chosen a column, you can use the position cursor to move up and down to the different data entries. The line at the bottom of the screen shows you what entry you are currently working on. For example, $L_3(4)$ would be the fourth entry in list L_3, and $L_1(19)$ would be the nineteenth entry in list L_1. Once you punch in a number, pressing $\boxed{\text{ENTER}}$ will enter the data into the list, and then the cursor will automatically move down the list to the next space.

Once the data is entered, press $\boxed{\text{STAT}}$ $\boxed{>}$ to scroll the cursor to the right to the **CALC** menu. Pressing $\boxed{1}$ gets the program called 1-Var Stats. Now enter the column your data is in as follows: If your data is in column L_1 then press $\boxed{\text{2nd}}$ $\boxed{1}$. The variable L_1 will appear on the screen as

1-Var Stats L_1 Now press $\boxed{\text{ENTER}}$ and you will see

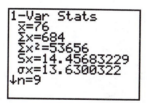

so the mean written as $\bar{x}$ is given as 76, and the standard deviation written as σ_x is 13.63, as in the last example. ■

Exercises 3.1

1) Let $x_1 = 7$, $x_2 = 5$, and $x_3 = -3$. Calculate $\sum x$.

2) Let $a_1 = 2$, $a_2 = 7$, $a_3 = -1.5$, $a_4 = 7$, and $a_5 = -3$. Calculate $\sum a$.

3) Let $y_i = \left(\dfrac{1}{2}\right)^i$ for $i = 1$ to 4. Calculate $\sum y$.

4) Let $x_i = 2^{i+1} + 1$ for $i = 1$ to 5. Calculate $\sum x$.

5) Your division at work has five employees. The hourly salaries for these five people are $7.45, $8.20, $6.50, $7.00, and $8.35. What is the mean hourly salary for your division?

6) The final score for a bowling team is defined as the average of the four individual scores. Team A bowled 134, 156, 122, and 189. Team B bowled 165, 174, 132, and 129. Who won the tournament?

7) Your Statistics class has 12 students in it. When questioned about how much time per week was spent on homework, the students answered; 2, 4, 3, 2½, 2, 3, 3½ , 4, 3, 1, 3½, and 2 hours. What is the mean amount of time spent on homework?

8) Suppose that the student scores, S, on the final exam in your history class are 64, 84, 52, 57, 89, 82, 79, 66, 71, 83 and 46. Compute the mean μ, and the standard deviation $\sigma = \sqrt{\dfrac{\sum (S-\mu)^2}{n}}$. If you wish, use your calculator.

9) Suppose that the weights W of the students in your gym class are 125, 182, 160, 111, 99, 210, 176, and 145. Compute the mean and standard deviation.

3.2 Weighted Mean

Sometimes when we need the mean of a set of quantities, some of these quantities may be more important than others. For example, the grade on the final exam usually counts more heavily than the grade on an in-class test. In that event, we need to use the notion of a <u>weighted mean</u>.

Definition: Given a list of quantities x_1, x_2, x_3, $\cdots$, x_n, and a corresponding list of weights, w_1, w_2, w_3, $\cdots$, w_n, we define the <u>weighted mean</u> to be

$$\frac{x_1 w_1 + x_2 w_2 + \cdots + x_n w_n}{w_1 + w_2 + \cdots + w_n},$$

or in Σ-notation:

$$\frac{\sum (x \cdot w)}{\sum w}.$$

Example 1:   In your Statistics course, there will be four in-class tests each weighted at $\frac{1}{6}$ of the course grade, and a final exam weighted at $\frac{1}{3}$ of the course grade. Your test results are 85, 93, 56, 87, and you got a 90 on the final. Compute your course grade.

Solution: Since the results are not weighted equally, we must use the weighted mean. So:

$$\frac{\sum (x \cdot w)}{\sum w} = \frac{85\left(\frac{1}{6}\right) + 93\left(\frac{1}{6}\right) + 56\left(\frac{1}{6}\right) + 87\left(\frac{1}{6}\right) + 90\left(\frac{1}{3}\right)}{\frac{1}{6} + \frac{1}{6} + \frac{1}{6} + \frac{1}{6} + \frac{1}{3}}$$

$$= 83.5 \quad \blacksquare$$

Example 2:   Last year, you made the following investments: $1000 in a CD at 5%, $500 in a risk fund that yielded 25%, and $200 in another risk fund that lost 20%. Find your average yield, in percent.

Solution: Your yields were 5%, 25% and −20%, but they are not weighted equally because you invested differing amounts. Notice that the percents are the quantities you are averaging, and the dollar amounts are the weights to give those percents. So:

$$x_1 = 5\% \qquad\qquad\qquad w_1 = 1000$$

$$x_2 = 25\% \qquad\qquad\qquad w_2 = 500$$

$$x_3 = -20\% \qquad\qquad\qquad w_3 = 200$$

and the weighted mean is:

$$\frac{\sum (x \cdot w)}{\sum w} = \frac{(5\%)(1000) + (25\%)(500) + (-20\%)(200)}{1000 + 500 + 200}$$

$$= \frac{50 + 125 - 40}{1700}$$

$$\cong .079 = 7.9\% \quad \blacksquare$$

Exercises 3.2

1) In your Biology course, there will be three in-class tests each weighted at $\frac{1}{5}$ of the course grade, and a final exam weighted at $\frac{2}{5}$ of the course grade. Your test results are 80, 77, 90, and you got an 82 on the final. Compute your course grade.

2) In your Statistics course, there will be four in-class tests each weighted at 15% of the course grade, and a final exam weighted at 40% of the course grade. Your test results are 65, 72, 81, 67, and you got a 77 on the final. Compute your course grade.

3) Suppose you made investments of $2000, $1500, and $4000 in three funds with interest rates of 5%, 7½%, and 4%, respectively. What was your average yield?

4) Suppose that in your class 5 students are age 17, 15 students are age 18, and 8 students are age 19. Compute the average age using the notion of mean, and then weighted mean. Which way is easier?

Chapter 4

Factorials, Permutations, and Combinations

4.1 Factorials

Factorials come up frequently in Probability and Statistics. In some contexts, you will see expressions like $1 \cdot 2 \cdot 3 \cdot 4$, which is of course the number 24, right? This can also be written in the form 4!, and is read "four factorial." Similarly $5! = 1 \cdot 2 \cdot 3 \cdot 4 \cdot 5 = (4!) \cdot 5$, and since $4! = 24$, $5! = 24 \cdot 5 = 120$. We can continue this: $6! = (5!) \cdot 6 = 720$ and $7! = (6!) \cdot 7 = 5040$. You can see that these little things called factorials get pretty large pretty quick.

Definition: Let n be a positive whole number. We define $n!$ to be $1 \cdot 2 \cdot 3 \cdot 4 \cdots (n-1) \cdot n$, and call it "$n$ factorial." We also define 0! to be 1.

As you may have figured out by now $n! = n \cdot (n-1)!$. This allows for interesting mathematics. The fun starts when you take quotients of factorials.

Example 1:   Express the following as whole numbers.

a) 8!   b) $\dfrac{10!}{9!}$ c) $\dfrac{1000!}{998!}$

Solution: a) $8! = (7!) \cdot 8 = 5040 \cdot 8 = 40{,}320$

b) For this one, we could calculate the top and bottom separately, and then divide, but what a drudge that would be! But notice $10! = (9!) \cdot 10$, and so

$$\frac{10!}{9!} = \frac{(9!) \cdot 10}{9!} = 10. \quad \text{We're done.}$$

c) $$\frac{1000!}{998!} = \frac{1000 \cdot 999 \cdot (998!)}{998!}$$

$$= 1000 \cdot 999 = 999{,}000 \quad \blacksquare$$

Note: Scientific calculators have a factorial key, so that you can easily compute $n!$. Simply enter the number n and hit the $\boxed{!}$ key. For those of you using the TI-83, to compute 5! you would press:

What's the largest factorial your calculator can compute? Experiment and see!

Exercises 4.1

1)   Simplify the following expressions.

   a) $\dfrac{12!}{11!}$ b) $\dfrac{6!}{8!}$ c) $\dfrac{7!}{5! \cdot 2!}$

2)   Use your calculator to compute the following, if possible.

   a) 9!   b) 12!   c) 50!   d) 100!

3)   Compute $\dfrac{n!}{k!(n-k)!}$ for $n = 6$, and $k = 1$.

4)   Compute $\dfrac{n!}{k!(n-k)!}$ for $n = 200$, and $k = 0$.

4.2 **Permutations and Combinations**

Suppose you have 10 children, and you are about to dish out ice cream to them one-by-one. In how many different ways could you line them up to receive their ice cream? Well, there are 10 ways of choosing the lucky first one, and for each of those 10 choices, you have 9 choices for the second spot. So the number of ways of lining up the first two is $10 \cdot 9 = 90$. The number of ways of picking the third child is 8, and hence the number of ways of picking the first 3 is $10 \cdot 9 \cdot 8 = 720$. So the number of ways of lining them all up is $10 \cdot 9 \cdot 8 \cdot 7 \cdot 6 \cdot 5 \cdot 4 \cdot 3 \cdot 2 \cdot 1 = 10!$ (This is why factorial notation is so useful.) A way of lining up or ordering a set of things is called a __permutation__ of those things. The idea can be generalized into a theorem.

Theorem A: The number of permutations of n things is $n!$

Now suppose you are in a class of 10 students, and you need to select a president, a secretary, and a treasurer. In how many ways can this be done? In light of the previous discussion, the answer is $10 \cdot 9 \cdot 8$. We can also write this as

$$10 \cdot 9 \cdot 8 = \frac{10 \cdot 9 \cdot 8 \cdot 7 \cdot 6 \cdot 5 \cdot 4 \cdot 3 \cdot 2 \cdot 1}{7 \cdot 6 \cdot 5 \cdot 4 \cdot 3 \cdot 2 \cdot 1} = \frac{10!}{7!} = \frac{10!}{(10-3)!}.$$

This really is the number of permutations of 10 things __taken 3 at a time__. This result can be generalized.

Theorem B: The number of permutations of n things taken r at a time is given by

$$\frac{n!}{(n-r)!} = n(n-1)(n-2)\cdots(n-(r-1)).$$

(This number is denoted $_nP_r$.)

Example 1: Fifteen people entered the poetry contest for a gold, a silver, and a bronze medal, and an honorable mention. How many possible outcomes are there?

Solution: This is the number of permutations of 15 things, 4 at a time, which according to Theorem B is

$$_{15}P_4 = \frac{15!}{(15-4)!} = \frac{15!}{11!} = 15\cdot14\cdot13\cdot12 = 32,760. \quad\blacksquare$$

Example 2: All the fraternities at a particular college have to have names consisting of 3 Greek letters, with no repetitions. There are 24 letters in the Greek alphabet. How many possible fraternity names are there?

Solution: This is equal to the number of ways to line up 24 letters, 3 at a time, which is the same as $_{24}P_3$. So there are

$$_{24}P_3 = \frac{24!}{21!} = 24\cdot23\cdot22 = 12,144$$

possible fraternity names. $\blacksquare$

There is one more twist coming. Suppose instead of an executive board of president, secretary, and treasurer, your class wants a council of 3 <u>equal</u> members. In how many ways can that council be chosen from the class of 10? (The point is that in a council, all members are of equal status, and so the council {Tom, Dick, Harry} is the same as the council {Harry, Dick, Tom}. Order does not matter.) That's the same as asking in how many ways you can combine 10 things, 3 at a time. This is denoted $_{10}C_3$.

<u>Definition:</u>   The number of combinations of n things taken r at a time is denoted $_{n}C_r$.

It is also denoted $\begin{pmatrix} n \\ r \end{pmatrix}$.

Example 3:   Answer the question raised above: In how many ways can a council of 3 be chosen from a group of 10 people?

Solution: The answer is denoted $_{10}C_3$, but that tells us nothing. Consider how many executive boards of president, secretary, and treasurer could be made from a class council of 3. From Theorem A, 3! is the answer. So for every class council, there are 3! executive boards. Hence the number of possible class councils, multiplied by 3!, gives the total number of possible executive boards: $_{10}C_3(3!) = {}_{10}P_3$.

Therefore $_{10}C_3 = \dfrac{{}_{10}P_3}{3!} = \left(\dfrac{10!}{7!}\right)\dfrac{1}{3!} = \dfrac{10!}{7!3!}$,

which equals $\dfrac{7! \cdot 8 \cdot 9 \cdot 10}{7!3 \cdot 2 \cdot 1} = \dfrac{720}{6} = 120.$ ∎

Remark: In this example, $_{10}C_3$ was equal to $\dfrac{10!}{7!3!}$. This pattern holds in general, as stated in the following theorem.

Theorem C: The number of combinations of n things taken r at a time is given by

$$\boxed{{_n}C_r = \frac{n!}{(n-r)!r!}.}$$

Example 4:   Twelve candidates are running for town council, but there are only 8 positions open. How many possible outcomes are there?

Solution: We're looking for the number of ways of choosing 8 things from a set of 12.   This is equal to the number of combinations of 12 things, 8 at a time, and is given by

$$_{12}C_8 = \frac{12!}{4!8!}$$

$$= \left(\frac{12!}{8!}\right)\left(\frac{1}{4!}\right)$$

$$= (12 \cdot 11 \cdot 10 \cdot 9)\left(\frac{1}{4 \cdot 3 \cdot 2 \cdot 1}\right) = 495. \quad \blacksquare$$

Remark: How do you tell a permutation from a combination? That is the question. It's easy if you recall that a permutation is a way of lining things up, so that <u>order matters</u>, as in the case of choosing an executive board. (For example, it makes a difference which of the three is president, etc.) In a combination, order doesn't matter, as in the example of the class council.

Example 5:   Over the years you've built up an esoteric collection of 100 baseball cards, all different, all unique. Little cousin Joey is visiting and is very impressed. Your mom says you've got to give him 5 of the cards. In how many ways can this be done?

Solution: We leave to you the methods of transfer, and concentrate on the question: In how many ways can 5 cards be chosen from 100? This is like choosing a class council (order does not matter). So the answer is

$$_{100}C_5 = \frac{100!}{95!\,5!}$$

$$= \left(\frac{100!}{95!}\right)\left(\frac{1}{5!}\right) = \frac{100\cdot99\cdot98\cdot97\cdot96}{5\cdot4\cdot3\cdot2\cdot1}$$

$$= 75{,}287{,}520 \text{ ways.} \blacksquare$$

Example 6:   In your dreams, you're a rock star and you need to decide what songs to play, and in what order, for your upcoming show at the Pair–O–Dice lounge. You've got time to play 4 songs, and you have a repertoire of 45 tunes to choose from. How many possible song lists can you come up with?

Solution: Since the order of songs is important, you've got $_{45}P_4$ ways, which equals

$$\frac{45!}{41!} = 45 \cdot 44 \cdot 43 \cdot 42 = 3{,}575{,}880 \text{ ways.}$$

Better start writing the list today! ■

Using the TI-83 to compute $_nP_r$ or $_nC_r$ is very similar to computing the factorial.

To compute $_{45}P_4$ you would press: 45 [MATH] [<] [2] 4 [ENTER]

To compute $_{100}C_5$ you would press: 100 [MATH] [<] [3] 5 [ENTER]

When using other statistical calculators to compute $_nP_r$ or $_nC_r$ you need to use a toggle switch to enter paired data. The toggle switch looks like $\boxed{x \, _\triangleleft^\triangleright \, y}$ or $\boxed{x \leftrightarrow y}$. To enter a data pair you press the first number, then press $\boxed{x \, _\triangleleft^\triangleright \, y}$, then the second number .

To compute $_{45}P_4$ you would press $45 \boxed{x \, {}^{\triangleright}_{\triangleleft} \, y} \, 4 \boxed{{}_{n}P_r}$. The result would then appear as 3,575,880.

To compute $_{100}C_5$ you would press $100 \boxed{x \, {}^{\triangleright}_{\triangleleft} \, y} \, 5 \boxed{{}_{n}C_r}$. Again you would obtain the solution 75,287,520 .

Exercises 4.2

1) Calculate: a) $_8P_4$ b) $_8C_4$

2) Calculate: a) $_9P_2$ b) $_9P_1$ c) $_9P_9$ d) $_9P_8$

3) Calculate :   a) $_{13}C_4$ b) $_{13}C_1$ c) $_{13}C_{13}$ d) $_{13}C_9$

4) Another way of writing $_nC_r$ is $\binom{n}{r}$. Find $\binom{100}{7} - \binom{100}{93}$.

5) Find the number of different five-digit numbers, with no repeating digits.

6) You've won "Bookworm of the Year," and can buy three books for $1, out of a supply of 10,000 books. How many possible outcomes are there?

7) In how many different ways can 48 books, out of a supply of 50 books, be ordered on a shelf?

8) What can you say about $_nC_r$ and $_nC_{n-r}$? (Hint: Play around. Try $n = 8$, and $r = 3$ for example. Also check out Exercise 2, above.)

Chapter 5

Graphs

5.1 Data and Their Representations

In statistics, it is often useful to display data in diagrams called graphs. "A picture is worth a thousand words." Large amounts of data can be understood more readily if represented in the proper visual fashion. There are many kinds of graphs. We will look at several of them here.

Example 1: The following table shows the price of a share of stock of B & M Inc. at the close of the trading day for the week of April 6 – 10, 1998. Take the data given and display it by means of a graph.

Day	Closing Price of B & M Stock
Monday	$ 6.75
Tuesday	$ 7.25
Wednesday	$ 8.50
Thursday	$ 9.75
Friday	$ 11.25

Solution: Here, we use a <u>bar graph</u> to represent the stock price data as a function of weekday.

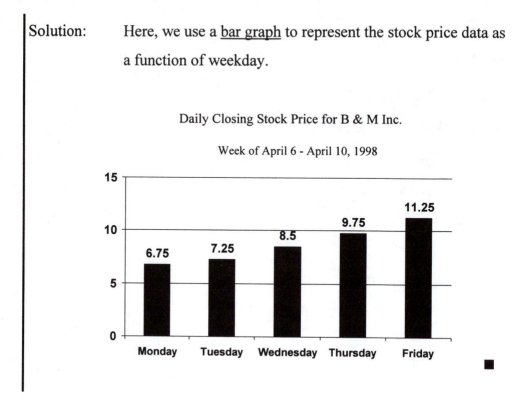

Daily Closing Stock Price for B & M Inc.

Week of April 6 - April 10, 1998

In a bar graph, the heights of the individual bars represent the value of what is called the <u>dependent variable</u> (in this case the stock prices). The labels under the bars indicate the "value" of the <u>independent variable</u>. Sometimes the labels are words, other times they are categories, numerical intervals, or numbers.

A <u>histogram</u> is a variation of a bar graph and it comes up in statistics when summarizing data. Here's an example:

Example 2:   Suppose you have measured the heights of all the students in your class and created the following table.

4'10" $\leq H <$ 5'	|	1
5' $\leq H <$ 5'2"	|||	3
5'2" $\leq H <$ 5'4"	|||	3
5'4" $\leq H <$ 5'6"	ЦЩ |	6
5'6" $\leq H <$ 5'8"	ЦЩ |||	8
5'8" $\leq H <$ 5'10"	ЦЩ	5
5'10" $\leq H <$ 6'	|	1
6' $\leq H <$ 6'2"	||	2
6'2" $\leq H <$ 6'4"	|	1

Display this data by means of a bar graph, showing the frequency of the heights in the various height intervals. In statistics, these intervals are called <u>classes</u>.

Solution: The independent data in this case are the heights, while the dependent data here are the frequencies of heights in each height interval.

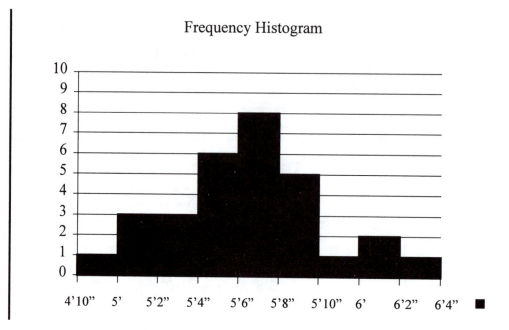

Frequency Histogram

Usually, the intervals are chosen equal in size. If so, the TI-83 can produce simple histograms from entered data. This can be a great way to get a quick representation, literally at the touch of a button. However, you need to know how to enter the data, and how to set any program parameters.

Example 3:   Consider the following final exam data taken from an Introduction to Nursing class:

71, 34, 72, 44, 52, 76, 54, 46, 57, 82, 63, 64, 91, 68, 66, 71, 73, 74, 85, 74, 56, 77, 78, 61, 84, 67, 86, 64, 87, and 92

Using your calculator, enter the data and plot a histogram using class intervals of width 10. That is, group the data into classes 1–10, 11–20, 21–30, 31–40, 41–50, 51–60, 61–70, 71–80, 81–90, and 91–100.

Solution: Example 4 on page 49 gives an outline of how to enter data into the TI-83. Assuming that the data has been entered into a data list, we proceed by first setting the window parameters as shown below. You can set window parameters by first pressing [WINDOW] and then using the scroll keys to position the cursor on the particular line you wish to edit, and then enter a new number. Since the exam results vary between 0 and 100, we will choose $X_{min} = 0$ and $X_{max} = 100$. Since y represents a count of the number of grades in a given interval we choose $Y_{min} = 0$. Since no one class contains more than 10 grades we also choose $Y_{max} = 10$. The class width on the TI-83, in this case 10, is set using the X_{scl} parameter. The window parameters we chose are show here.

```
WINDOW
 Xmin=0
 Xmax=100
 Xscl=10
 Ymin=0
 Ymax=10
 Yscl=1
 Xres=1
```

Set your window parameters to match ours now. To create a histogram on the TI-83, you first enter the Stat Plot routine by pressing [2nd] [Y=] . Your screen will look like the following.

```
STAT PLOTS
1:Plot1...Off
   L1   L2
2:Plot2...Off
   L1   L2
3:Plot3...Off
   L1   L2
4↓PlotsOff
```

The TI-83 has the ability to create and store 3 different Stat plots named Plot1, Plot2, and Plot3. Pressing [ENTER] again will edit Plot 1 and you will see the following screen:

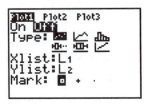

The cursor will be flashing over the word On, and the word Off will be highlighted in black to indicate that the plot is turned off. Pressing [ENTER] again will switch the plot on. There are six types of statistics plots on the TI-83. There are little pictures of them on the Type: line. To choose the histogram, simply select the top right-most graph by scrolling down one, right two, and then press [ENTER]. The screen will now look like:

Scroll down to Xlist: and select the correct data list, that is, L₁, L₂, etc. Press [GRAPH] and you will see:

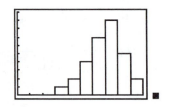

Example 4:   Here is a recording of the temperature in Times Square, N.Y.C. between 8 AM and 8 PM on a certain day. (This kind of graph is called a <u>line graph</u>.)

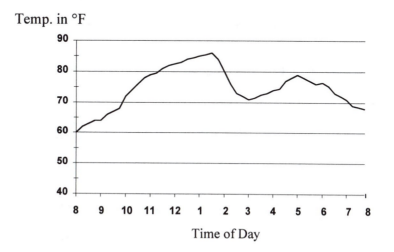

Temp. in °F

Time of Day

a) Approximately, what was the temperature at noon?

b) What was the highest temperature, and when did it occur?

c) What was the lowest, and when did it occur?

d) Give a rough estimate of the average temperature over that 12-hour interval.

Solution: a) At noon, the temperature was roughly 83° F.

b) The highest temperature was about 86° F at 1:30 PM.

c) The lowest temperature was about 60° F, at 8 AM.

d) Looking at the graph, we see that the average was less than 80° F and greater than 70° F, maybe around 75° F. ■

Remark: In a line graph, every point on the graph is represented by a pair of numbers, given in the following way: (independent variable, dependent variable). In Example 4, the independent variable is time and the dependent variable is the temperature. So the point on the curve that answers the first question above is the point (12,83). Data of this sort are called paired data.

So we see that paired data can be given by tables and by graphs. They can also be given by words and by symbolic expressions. It is important that you can easily change data representations from one form to another. Consider the next example.

Example 5:   Hurts Car Rental charges $50 for a weekend, plus 15 cents per mile driven. Let C be the rental cost, in dollars, and let d be the distance driven, in miles. a) Express C in terms of d using an equation. b) Make a line graph, showing C as the dependent variable and d as the independent variable.

Solution: a) $C = 50 + $ (number of miles driven)(0.15);

so $C = 50 + (0.15)(d)$.

b) To find the graph of C in terms of d, it is useful to make a table of values of d, and the corresponding values of C, like the following.

d (in miles)	C (in dollars)
0	50
50	57.50
100	65
200	80
300	95
500	125
1000	200

Each line of the table yields a point in the graph. For example, the first line in the table goes to the point (0,50) in the line graph. Placing the point is called <u>plotting</u> the point. By plotting all of the points in this table we obtain the following.

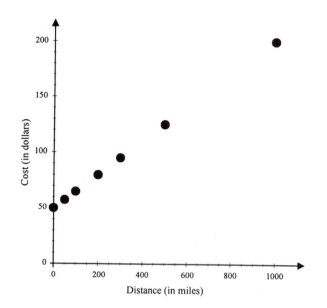

This gives you an incomplete graph, of course, because what if you drive 700 miles or 2000 miles? Well, you can plot those points too. Obviously you'll never finish the complete graph by only plotting points. We'll see in Chapter 6 that $C = 50 + (0.15)(d)$ is the equation of a line. So we know that it's enough to plot any two points and draw a straight line through them. From the data, choose, for example, $(0,50)$ and $(1000,200)$. The following graph is the result.

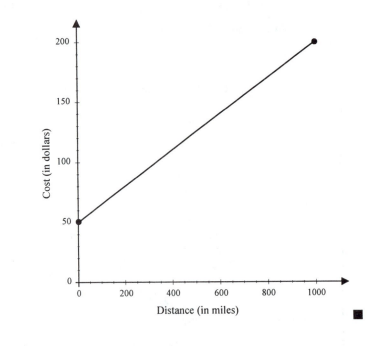

The graph of a set of paired data points is also called a <u>scatter diagram</u>. The TI-83 can be used to generate scatter diagrams.

Example 6:   Consider the following height and weight data collected from a health study.

Height (in inches)	Weight (in pounds)
71.3	146
68.5	165
67.4	157
62.6	128
68.1	155
67.0	190
70.0	147
65.4	144
61.6	144
73.1	172
69.8	135
64.1	152
72.0	198
66.8	133

Using your calculator, enter the data and plot a scatter diagram.

Solution: Example 4 on page 49 gives an outline of how to enter data into the TI-83. Assuming that the data has been entered into data lists L1 and L2, we start off like we did in Example 3 of this section. First choose which plot you wish to use; we chose Plot1 again. Pressing [2nd] [Y=] [ENTER] will enter the Plot1 editor. Make sure the plot is toggled to the "on" position, and then scroll down to the Type: line.

Select the top left-most graph. Next, select the correct data lists. We entered the height data in list L1, and the weight data into list L2, so our plot parameter screen looked like:

To get the scatter plot, press [ZOOM] [9]. Your calculator will display something like the following graph on the left. (Press [WINDOW] to get the display on the right.)

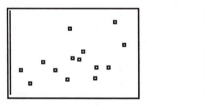

The parameters Xscl and Yscl are left the same as the last time you used your calculator, so they may not be the same as the ones shown above, and they may not be appropriate for the problem. If you change them to 2 and 10 respectively, and then press [GRAPH] you will see:

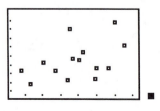

Exercises 5.1

1) For your Sociology 101 Course, you poll 455 homes in your neighborhood, asking each family how many Barbie dolls they own. You create the following table:

Number of Barbies	Number of Families
0 – 10	122
11 – 20	220
21 – 30	81
31 – 40	21
41 – 50	11

Display this data by means of a bar graph.

2) Consider the following age data collected from 40 students in an Introduction to Management course:

18, 19, 20, 30, 18, 20, 19, 24, 19, 18, 22, 27, 25, 18, 19, 18, 23, 27, 18, 19,

18, 17, 28, 31, 18, 19, 20, 19, 19, 17,21, 23, 18, 20, 25, 28, 33, 17, 22, 18.

Group the data into classes 17–19, 20–22, 23–25, 26–28, 29–31, and 32–34. Using your calculator, enter the data and plot a histogram for this data.

3) The daily volume, or total number of shares, traded on the stock exchange is given in the graph below, for a six-week period during April and May of 1998.

VOLUME (in millions of shares)

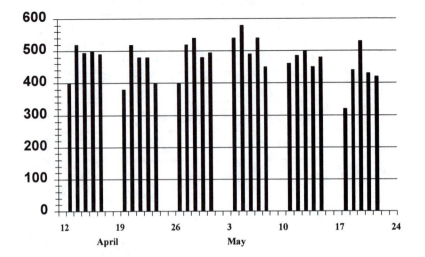

a)   Approximately how many millions of shares were traded on Monday, May 4th? (Note: Dates shown on the axis are Sundays, so that May 4th is the first bar after the number 3.)

b)   What was the largest volume and when did it occur?

c)   What was the smallest volume, and when did it occur?

d)   What was the average volume for the week of April 26th?

e)   Convert this graph for the week of April 19 to a table.

4) Your local Pizza Parlor sells a large pizza for $6 plus $1.50 for each topping requested.

  a) Let C be the cost of a pizza, and let n be the number of toppings. Write down a symbolic expression that gives C in terms of n.

  b) Make a table to represent the cost of a pizza for various numbers of toppings from 0 to 5.

  c) Use a bar graph to represent the pizza cost in terms of the number of toppings.

5) Consider the following data relating the height and weight of 15 people in your gym class. Use your calculator to plot a scatter diagram representing the data.

Height (in inches)	Weight (in pounds)
61.3	126
69.5	185
62.4	127
67.6	158
68.1	190
64.0	155
71.0	167
63.4	154
61.8	148
72.1	172
67.8	145
64.5	158
68.2	178
72.1	197
65.8	133

Chapter 6

Lines

6.1 Lines and Their Equations

Straight lines are the simplest of all curves, and one of the central ideas in statistics is to use lines to approximate complicated real-life data. So you've got to be proficient at lines and their equations. Lines have various degrees of steepness, which in mathematics is called <u>slope</u>. Consider Figure 1.

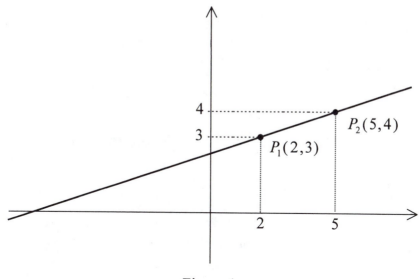

Figure 1

As you go from point P_1 to point P_2, you increase vertically a distance of 1 (called the <u>rise</u>), and horizontally a distance of 3 (called the <u>run</u>). The slope of line L is defined as $\dfrac{\text{rise}}{\text{run}} = \dfrac{1}{3}$. Notice the notation we use: $P_1(2,3)$ stands for the point P_1, whose coordinates are $(2,3)$.

Definition: Suppose the line L contains the two points $P_1(x_1, y_1)$ and $P_2(x_2, y_2)$, as shown in Figure 2.

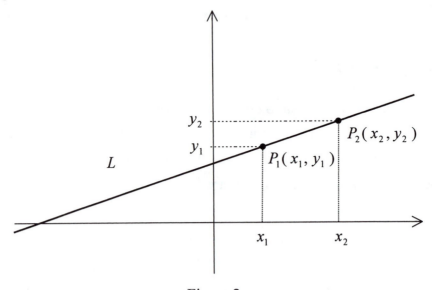

Figure 2

Assume L is not vertical, and hence that $x_2 - x_1 \neq 0$. The ratio $\dfrac{y_2 - y_1}{x_2 - x_1}$ is called the slope of L, and is usually represented by the letter m. So:

$$\boxed{\text{Slope} = m = \frac{y_2 - y_1}{x_2 - x_1}}$$

<u>Remarks:</u>

1) Notice that if $x_2 - x_1 = 0$, then $x_2 = x_1$, the line is vertical, and the ratio

$\dfrac{y_2 - y_1}{x_2 - x_1}$ makes no sense. That's why we have defined the notion of slope only

for nonvertical lines.

2) Notice also what happens if we had chosen not P_1 and P_2, but two other points on the line, say P_3 and P_4, to calculate the slope. By using similarity of triangles, we could prove that the slope calculated using P_1 and P_2 would equal the slope calculated using P_3 and P_4. So any pair of points will do to calculate the slope.

3) If, as we go from left to right, the line rises, then the slope is a positive number. On the other hand, if the line falls, the slope is negative.

Consider the equation $y = \frac{1}{3}x + 2$. Now think of all points of the form (x,y) where it is true that $y = \frac{1}{3}x + 2$. Well, $(0,2)$ is such a point, as well as $\left(1,\frac{7}{3}\right)$, $\left(2,\frac{8}{3}\right)$, $(3,3)$, $\left(-1,\frac{5}{3}\right)$, $\left(-2,\frac{4}{3}\right)$, and $(-3,1)$. Let's see where these points are (that's called <u>plotting</u> them) in Figure 3.

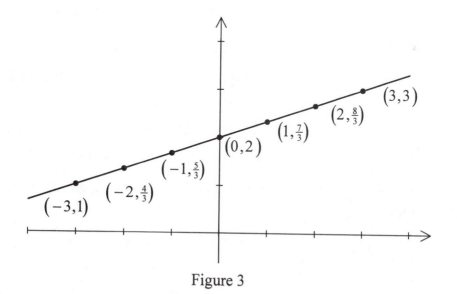

Figure 3

They all lie in the same straight line. This line is said to be the graph of the equation $y = \frac{1}{3}x + 2$. Using two of the points, we can calculate that $m = \frac{1}{3}$. Notice that the coefficient of x is also $\frac{1}{3}$. This is not a coincidence. The height at which the line crosses the y-axis is called the y-intercept, which is 2 in this example. Sometimes the point of intersection, (0,2), is also called the y-intercept.

More generally:

$$y = mx + b \text{ is the equation of the line of slope } m \text{ and } y\text{-intercept } b.$$

Sometimes the equation of the line will be given as $y = ax + b$, and sometimes as $y = \beta_0 + \beta_1 x$. In any case just remember that the slope is the coefficient of the x-term (the number that is multiplied by x). So if the line is given as $y = ax + b$, then b is the y-intercept and a is the slope. If the line is given as $y = \beta_0 + \beta_1 x$, then β_0 is the y-intercept and β_1 is the slope.

Example 1:   Consider the line $y = 1.37 + 2.47x$. What is the slope and the y-intercept of this line?

Solution: The coefficient of the x term is 2.47, so that is the slope. The other number, 1.37, is the y-intercept. ∎

Often the variables will be letters other than x and y (typically corresponding to physical variables).

Example 2:   Consider the line $H = 37.68 - 12.18W$. What is the slope and the H-intercept of this line?

Solution: In this case the variables are H and W. H is playing the role of y, and W is playing the role of x. The number -12.18 is the coefficient of the variable W and so the slope of this line is -12.18. (Don't forget the negative sign!)

The number 37.68 is the y-intercept. ■

Example 3:   Graph the line $y = 2x - 5$, labeling two points.

Solution: The line $y = 2x - 5$ has slope 2, and goes through $(0,-5)$. We can get a second point on the line by starting at $(0,-5)$ and moving to the right 1 unit, and up 2 units (since $m = 2$), because then we'll have $\dfrac{\text{rise}}{\text{run}} = \dfrac{2}{1}$ as we should. [The trick is always: go right 1, go up m.] The second point is $(1,-3)$. Two points determine the whole line. Using these two points we obtain the following graph.

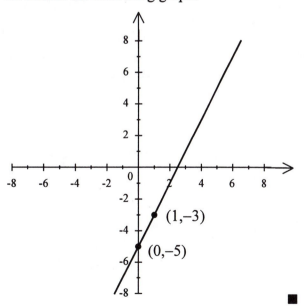

■

Example 4:   Graph the line $y = a + bx$, where $a = 2.37$ and

$b = -1.67$.

Solution: The line $y = 2.37 - 1.67x$, has slope -1.67 going through

$(0, 2.37)$. If we start from $(0, 2.37)$ and go over 1 and up m

(in this case $m = -1.67$, so we go down a distance of 1.67),

we get a second point on the line $(1, 0.7)$, and hence we

know the whole graph. Using these two points we obtain

the following graph.

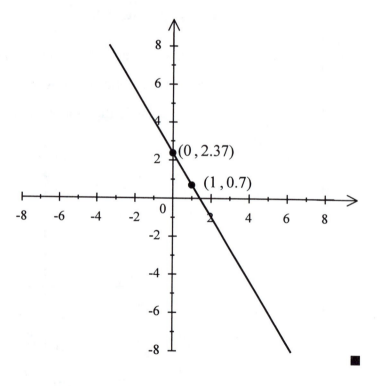

The TI-83 can produce regression lines from entered data. This can be a great way to get quick information.

Example 5:   Consider the height and weight data from Example 6 in Chapter 5 on page 77.

Using your calculator, enter the data and then compute and plot the regression line.

Solution: Example 6 on page 77 gives an outline of how to plot scatter diagrams of paired data using the TI-83. Assuming that the data has been entered into data lists L_1 and L_2 and a scatter plot is obtained, one can then calculate and graph a regression line as follows.

Pressing [STAT] [>] [4]   gets the program LinReg(ax+b). Now press

[2nd] [1] [,] [2nd] [2] [,] [VARS] [>] [1] [ENTER]

and you will see:

```
LinReg(ax+b) L1,
L2,Y1█
```

Now press ENTER and you will see:

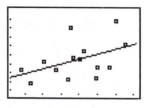

Note the slope, a, and intercept, b, are given here.

Now press ZOOM 9 and you will see:

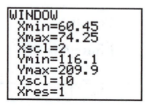

Pressing WINDOW displays the window as shown here:

```
WINDOW
 Xmin=60.45
 Xmax=74.25
 Xscl=2
 Ymin=116.1
 Ymax=209.9
 Yscl=10
 Xres=1
```

These values allow you to label the scatter plot axes appropriately. Just like on page 78, the values for the axis scales are left the same as the last values, so your window may differ from ours. To get ours, reset the two parameters to 2 and 10 respectively. ■

Exercises 6.1

In Exercises 1–5, graph the function by plotting two points each. (Why is it enough to plot only two points ?)

1) $f(x) = \dfrac{1}{3} x$ 2) $y = -1.5\ x$

3) $g(x) = .6\ x$ 4) $f(x) = 5\ x$

5) $y = 1000\ x$

6) Given the line $y = 4.56 + 5.23\ x$, what is the slope and the y-intercept? Graph the line.

7) Given the line $S = 1.28 - 15.47\ T$, what is the slope and the S-intercept? Graph the line.

8) Graph the line $y = -3.154 + 2.65\ x$.

9) Graph the line $H = 15.87 + 4.23\ W$.

10) Graph the line given by the equation $Q = -1.28 + 5.12\ B$.

11) Graph the line given by the equation $C = -2.98 - 3.43\ D$.

12) Using your TI-83, calculate and plot the regression line for the data from Chapter 5 Exercise 5 on page 81.

Chapter 7

Solving Equations and Inequalities

7.1 Solving Equations

There are several situations in statistics where you will have to solve an equation for one specific variable in terms of other given variables. For example, when converting from nonstandard to standard normal distributions, the equation

$$z = \frac{x - \mu}{\sigma}$$

is used, where μ is the mean and σ is the standard deviation. Sometimes you will have to solve this equation for x in terms of the remaining variables. Another example is when considering the margin of error formula:

$$E = z_{\alpha/2} \cdot \frac{\sigma}{\sqrt{n}}$$

It will happen that you need to solve this equation for n in terms of E, $z_{\alpha/2}$, and σ. In any case, once you get over the strange symbols in these equations, and realize that they just represent numbers, it turns out that equations of this sort are easy to solve. Let's start by considering simple cases.

Example 1: Solve for x: $4x - 16 = 0$

Solution: First, get rid of the -16 by adding 16 to both sides:

$$4x - 16 + 16 \;=\; 0 + 16,$$

which gives $\qquad\qquad 4x \;=\; 16.$

Now, get rid of the 4 in front of the x by dividing both sides by 4:

$$\frac{4x}{4} \;=\; \frac{16}{4},$$

giving $\qquad\qquad x \;=\; 4.$ Easy!

It's always a good idea to check your answer by substitution. That is, if we substitute $x = 4$ back into the original equation, it works! It is certainly true that

$$4 \cdot 4 - 16 \;=\; 0. \;\blacksquare$$

Example 2:   Solve for x: $\qquad \dfrac{2}{3}x + 1 \;=\; 0$

Solution: Get rid of the 1 by subtracting it from both sides. This gives

$$\frac{2}{3}x + 1 - 1 \;=\; 0 - 1,$$

or
$$\frac{2}{3}x = -1.$$

Now, to isolate the variable x, divide by $\frac{2}{3}$.

To do this, just like in Chapter 1, multiply by its reciprocal, $\frac{3}{2}$:

$$\frac{3}{2} \cdot \frac{2}{3}x = \frac{3}{2} \cdot (-1)$$

This gives
$$x = -\frac{3}{2}. \quad \blacksquare$$

Example 3:   If $z = \frac{1}{2}$, $\mu = 10$, and $\sigma = 2$, the equation $z = \frac{x - \mu}{\sigma}$ becomes

$$\frac{1}{2} = \frac{x - 10}{2}.$$

Solve this equation for x.

Solution: First, rewrite this as

$$\frac{x - 10}{2} = \frac{1}{2},$$

and then multiply both sides of the equation by 2. This gives us

$$x - 10 = 1.$$

Now adding 10 to both sides of the equation gives us

$$x = 11. \blacksquare$$

Example 4:   If $z = 0.96$, $\mu = 1.23$, and $\sigma = 0.82$, the equation

$z = \dfrac{x - \mu}{\sigma}$ becomes

$$\frac{x - 1.23}{0.82} = 0.96.$$

Solve this equation for x.

Solution: First, multiply both sides by 0.82 giving

$$0.82 \cdot \frac{x - 1.23}{0.82} = (0.82) \cdot (0.96),$$

or $x - 1.23 = 0.7872$.

Now add 1.23 to both sides to get

$$x = 2.0172. \blacksquare$$

In each of the preceding examples we obtained a solution by "peeling the onion." The steps taken in stripping away the layers to expose the desired variable involved addition, subtraction, multiplication, or division by real numbers. Things can get a little more confusing when the equation you are solving involves more than the one variable, but the idea is exactly the same. So, keep in mind what you learned, and forge ahead!

Example 5:   Solve for x: $y^2 x + w^2 = 0$

Solution: Now the equation has several variables. Remember the variable you wish to solve for and concentrate on isolating that one variable. Here, we wish to solve for x. Even though things seem a little more complicated, x still appears in only one place, so you can start peeling. In this case, first peel away the w^2 by subtracting it from both sides to get

$$y^2 x \ = \ 0 - w^2,$$

or $$y^2 x = -w^2.$$

Next peel off the y^2 by division:

$$\frac{y^2 x}{y^2} = \frac{-w^2}{y^2},$$

resulting in

$$x = \frac{-w^2}{y^2}. \blacksquare$$

Again, we solve the equation by stripping away the layers to expose the desired variable. Unlike the first three examples, the last example also had variables, w and y, in addition to the variable we were trying to solve for, x. These variables, however, just stand for real numbers, and so the mathematical steps taken to peel them away are exactly the same as those taken in the other examples.

Example 6:   Given the margin of error formula: $E = z_{\alpha/2} \cdot \dfrac{\sigma}{\sqrt{n}}$.

Suppose $z_{\alpha/2} = 0.64$ and $\sigma = 0.18$. What must n be so that $E = 0.0576$?

Solution: Here, substitute the known values of E, $z_{\alpha/2}$, and σ into the equation to get

$$0.0576 = (0.64) \cdot \frac{(0.18)}{\sqrt{n}}.$$

Solve for n by first multiplying both sides of the equation by $\sqrt{n}$, to get

$$0.0576 \cdot \sqrt{n} = (0.64) \cdot \frac{0.18}{\sqrt{n}} \cdot \sqrt{n},$$

or $0.0576 \cdot \sqrt{n} = (0.64) \cdot (0.18)$.

Then, divide both sides by 0.0576 to get

$$\sqrt{n} = \frac{(0.64) \cdot (0.18)}{0.0576},$$

which is equivalent to

$$\sqrt{n} = 2.$$

Squaring both sides of the equation gives the result

$$n = 4. \quad \blacksquare$$

Example 7:   Given the margin of error formula: $E = z_{\alpha/2} \cdot \dfrac{\sigma}{\sqrt{n}}$. Solve this equation for n.

Solution: Follow the lead of the steps taken in the last example by first multiplying both sides of the equation by $\sqrt{n}$, to get

$$E \cdot \sqrt{n} = z_{\alpha/2} \cdot \frac{\sigma}{\sqrt{n}} \cdot \sqrt{n},$$

or $E \cdot \sqrt{n} = z_{\alpha/2} \cdot \sigma$.

Now, divide by E to get

$$\sqrt{n} = \frac{z_{\alpha/2} \cdot \sigma}{E}.$$

Now, square both sides to get

$$n = \left[\frac{z_{\alpha/2} \cdot \sigma}{E} \right]^2 . \quad \blacksquare$$

Sometimes the variable you wish to solve for occurs more than once. In many cases this is not a problem, because you can rewrite the equation in such a way that the variable occurs only once, and then solve as usual.

Example 8:   Solve for x: $5x + 3 = 2x + 1$

Solution: Here x occurs more than once, but if we subtract $2x$ from both sides, we get

$$5x - 2x + 3 = 1$$

or $3x + 3 = 1.$

Notice that subtracting $2x$ from both sides is the same as moving the $2x$ across the equal sign where it shows up as $-2x$. Now, subtract 3 from both sides to get

$$3x = -2,$$

where all the terms <u>not</u> containing x are now moved to the right side.

Dividing by 3 gives the final answer

$$x = \frac{-2}{3}. \quad \blacksquare$$

Example 9:   Solve for x: $2x + 5y = 3x + y + 1$

Solution: Even though this appears more complicated because of the extra y terms floating around, it really isn't any more difficult. Here, the x occurs more than once, so subtract $3x$ from both sides, to get

$$-x + 5y = y + 1.$$

Notice that the terms containing x are all on the left side.

Now, we subtract $5y$ from both sides:

$$-x = y + 1 - 5y,$$

which gives $\qquad -x = 1 - 4y.$

(Notice that the terms not containing x are all on the right side.)

Dividing by -1 gives the final answer:

$$x \; = \; \frac{(1 - 4y)}{-1}$$

$$= \; \frac{-(1 - 4y)}{1}$$

$$= \; -1 + 4y \quad \blacksquare$$

Exercises 7.1

1) Solve for x: $3x - 4 = 0$

2) Solve for x: $\dfrac{5}{12}x - 35 = 0$

3) Solve for z: $\dfrac{4}{3}z - 1 = \dfrac{1}{10}$

4) Solve for x: $z = \dfrac{x - \mu}{\sigma}$

5)   Given the margin of error formula: $E = z_{\alpha/2} \cdot \dfrac{\sigma}{\sqrt{n}}$, and that $z_{\alpha/2} = 0.624$ and

$\sigma = 0.15$. What must n be so that $E = 0.00585$?

6)   Given the margin of error formula: $E = t_{\alpha/2} \cdot \dfrac{s}{\sqrt{n}}$. Solve for n in terms of the

remaining variables E, $t_{\alpha/2}$, and s.

7)   Solve the equation $E = z_{\alpha/2} \cdot \sqrt{\dfrac{pq}{n}}$ for n in terms of the remaining variables.

8)   Solve for y: $\dfrac{1}{2}y - \dfrac{1}{3} = \dfrac{1}{6} - 2y$

9)   Solve for x: $3y^2 x + z^2 = 0$

10)   Solve for y: $2zy + 3z + 1 = 0$

7.2 <u>Quadratic Equations</u> (in case you ever need them)

The equation $ax^2 + bx + c = 0$ is called a <u>quadratic equation,</u> because "quadratum" is Latin for "square". Notice that x occurs in two places, so you can't use the method of "peeling the onion" if you want to solve for x. However, it is shown in algebra texts that the solutions are of the form

$$x = \frac{-b \pm \sqrt{b^2 - 4ac}}{2a}.$$

This is the quadratic formula. It gives you the solutions of that equation by simply substituting in the values of a, b, and c. No fuss, no bother, no completing the squares, no messy onion peels. It is one of the most useful formulas you will meet. How many solutions are there? Well, it depends. Sometimes one, sometimes two, sometimes none.

Remark: The "±" sign in the quadratic formula means that there are possibly two solutions, one if we pick the "+" sign and the other if we pick the "−" sign. Sometimes the two solutions are, in fact, equal. (When does this happen?)

> Example 1:   Solve for x: $x^2 - 3x + 2 = 0$
>
> Solution: Here $a = 1$, $b = -3$, and $c = 2$. So, according to the quadratic formula

$$x = \frac{-(-3) \pm \sqrt{(-3)^2 - 4 \cdot 1 \cdot 2}}{2 \cdot 1},$$

which simplifies to

$$x = \frac{3 \pm \sqrt{9 - 8}}{2} = \frac{3 \pm 1}{2}.$$

When the "+" sign is used we get the solution

$$x = \frac{3 + 1}{2} = \frac{4}{2} = 2.$$

When the "−" sign is used we get the solution

$$x = \frac{3 - 1}{2} = \frac{2}{2} = 1.$$

So there are two solutions: $x = 1$ and $x = 2$. Can you see how useful this formula is? ■

Example 2:   Solve for x: $2x^2 - 3x + 1 = 0$

Solution: Here $a = 2$, $b = -3$, and $c = 1$, and $b^2 - 4ac = 1$. So, according to the quadratic formula

$$x = \frac{-(-3) \pm \sqrt{(-3)^2 - 4 \cdot 2 \cdot 1}}{2 \cdot 2},$$

which simplifies to

$$x = \frac{3 \pm \sqrt{9 - 8}}{4} = \frac{3 \pm 1}{4}.$$

The "+" sign gives $\qquad x = \dfrac{3 + 1}{4} = \dfrac{4}{4} = 1.$

The "−" sign gives $\qquad x = \dfrac{3 - 1}{4} = \dfrac{2}{4} = \dfrac{1}{2}.$

So there are two solutions: $x = 1$ and $x = \dfrac{1}{2}$. ■

<u>Remark:</u> What happens in the quadratic formula when $b^2 - 4ac = 0$? Well, then

$$x = \frac{-b \pm \sqrt{0}}{2a} = \frac{-b}{2a}$$: so, only one solution! What if $b^2 - 4ac < 0$? Then

$\sqrt{b^2 - 4ac}$ is not a real number, and the equation has no real solutions.

Sometimes you can solve a quadratic equation by "peeling the onion" more quickly than using the quadratic formula. This happens when the coefficient $b = 0$. Consider the next example.

Example 3:   Solve for x: $x^2 - 9 = 0$

Solution: Here $a = 1$, $b = 0$, and $c = -9$. Instead of using the quadratic formula we will "peel the onion". First we add 9 to both sides of the equation to get

$$x^2 = 9.$$

Now we take square roots of both sides of the equation, remembering that all positive numbers have two square roots. So

$$\pm x = \pm 3.$$

This means $x = 3$ or $x = -3$. ∎

Exercises 7.2   Find the solutions for Exercises 1–10.

1) $x^2 + 5x + 4 = 0$

2) $g^2 - 2g - 8 = 0$

3) $x^2 + 6x + 9 = 0$

4) $2y^2 + y - 1 = 0$

5) $y^2 - 8y + 16 = 0$

6) $\dfrac{x^2}{4} + 2x + 1 = 0$

7) $3y^2 - 4 = 0$

8) $x^2 + 3x = 4$

9) $x^2 + 7x + 4 = 0$

10) $3y^2 + 1.8y - 1.2 = 0$

7.3 <u>Simple and Compound Inequalities</u>

Just as there are equations in x (or y or z, etc.) that we sometimes wish to solve, so there are inequalities in x that we may wish to solve, meaning to find all values of x for which the inequality is true. When solving equations, we add, subtract, multiply, divide, square, etc. both sides of the equation. We shall do something similar for inequalities, although you should know: the rules are a little bit different. Here they are.

1. If $a < b$, and if c is any number, then

$$a + c < b + c$$

and $\qquad\qquad a - c < b - c.$

2. If $a < b$, and if c is any <u>positive</u> number, then

$$a \cdot c < b \cdot c$$

and $\qquad\qquad \dfrac{a}{c} < \dfrac{b}{c}.$

3. If $a < b$, and if c is any <u>negative</u> number, then

$$a \cdot c > b \cdot c$$

and $\qquad\qquad \dfrac{a}{c} > \dfrac{b}{c}.$

(Notice that the direction of the inequalities is reversed if c is negative.)

4. If $a < b$, and a and b are positive, then

$$\frac{1}{a} > \frac{1}{b}.$$

5. If $a < b$, and a and b are positive, then

$$a^2 < b^2$$

and
$$\sqrt{a} < \sqrt{b}.$$

To summarize these rules: As far as addition or subtraction of numbers to each side of the inequality is concerned, we can manipulate inequalities the same way we manipulate equations. It's when we multiply or divide both sides of the inequality by numbers that things get wiggy.

Just remember: As long as you are multiplying or dividing by positive numbers, the inequality keeps its direction. When we multiply or divide both sides of an inequality by a negative number we must reverse the direction of the inequality.

Example 1:   Solve for x: $5 - 3x > 17$

Solution: If this were an equation, we'd get rid of the 5 and the -3 to get to x. We'll do the same here, keeping in mind the above rules about the direction of the inequality.

First, we subtract 5 from both sides to get

$$-3x > 12.$$

Dividing by −3 gives us

$$x < -4.$$

(Note the change of direction.)

That's it! All numbers less than −4 are solutions. Hence, the set of solutions is $(-\infty, -4)$. ∎

Example 2: Solve for x: $7x + 3 \le 2x + 8$

Solution: Subtract 3 from both sides and get

$$7x \le 2x + 5.$$

Now subtract $2x$ to get

$$5x \le 5.$$

Dividing by 5 gives

$$x \le 1.$$

So the solution set is $(-\infty, 1]$. ∎

Example 3:   Solve for x: $3x - 8 \leq 5x - 2$

Solution: Add 8 to both sides and get

$$3x \leq 5x + 6.$$

Now subtract $5x$ to get

$$-2x \leq 6.$$

Dividing by -2 gives us

$$x \geq -3.$$

So the solution set is $[-3, \infty)$. ∎

All of the preceding examples illustrate inequalities that are called simple inequalities. If $a < b$ and $b < c$, we can combine these simple inequalities into the compound inequality $a < b < c$. Solving such inequalities is no tougher than simple inequalities.

Example 4:   Solve for x: $-9 < 3x - 7 \leq 8$

Solution: Add 7 and get

$$-2 < 3x \leq 15.$$

Now, dividing by 3 we get

$$-\frac{2}{3} < x \leq 5,$$

and so the solution set is $\left(-\dfrac{2}{3}, 5\right]$. ■

Example 5: Solve for x: $\qquad -5 < \dfrac{7 + 2x}{3} < 11$

Solution: To isolate the variable x we first need to multiply the inequality by 3 (the direction of the inequalities remain unchanged). We get

$$-15 < 7 + 2x < 33.$$

Now, subtract 7 from the inequality:

$$-22 < 2x < 26$$

Finally, dividing by 2 we obtain the solution:

$$-11 < x < 13 \quad ■$$

Example 6: Solve for σ: $\qquad 9 < \sigma^2 < 25 \qquad$ (Assume $\sigma > 0$.)

Solution: To isolate the variable σ we need to take the square root of each of the expressions:

$$\sqrt{9} < \sqrt{\sigma^2} < \sqrt{25},$$

and since $\sqrt{\sigma^2} = \sigma$ for $\sigma > 0$, we obtain

$$3 < \sigma < 5.$$

Note: In general, the solution to $a < \sigma^2 < b$ is

$$\sqrt{a} < \sigma < \sqrt{b}. \quad \blacksquare$$

Sometimes we need to multiply a compound inequality by -1. This can happen when computing so-called confidence intervals for differences of variables. Consider the following.

Example 7:   Solve for $b - a$: $-3 < a - b < 5$

Solution: Here we are asked to solve for $b - a$, which, although it has two parts, we can still regard as a single variable. Notice that $b - a = (-1)(a - b)$. So, to solve for $b - a$, we need to multiply this inequality by -1. Remember that the inequality signs change direction since $-1 < 0$, so we get

$$(-1)(-3) > (-1)(a - b) > (-1)5$$

which simplifies to

$$3 > b - a > -5.$$

Since $b-a > -5$ is the same as $-5 < b-a$, and $3 > b-a$ is the same as $b-a < 3$ we reorder the above inequality to look like

$$-5 < b-a < 3.$$

It's preferred to use the "<" form because then the numbers are ordered like they are on the number line. ∎

Exercises 7.3   In Exercises 1–12 solve the inequality.

1) $4x + 8 > 56$

2) $6x - 10 < 4$

3) $-2x + 3 \geq 7$

4) $3 - 9x \leq 30$

5) $1.3x - 5.78 > 3.24$

6) $3.47 - 4.3x < -2.7$

7) $3 < 5x - 1 < 28$

8) $-3 \leq 4x + 8 < 9$

9) $6 < 2 - 4x \leq 10$

10) $1.2 < 2x - 4.5 < 2.6$

11) $-0.99 \leq \dfrac{2.3x + 1}{0.34} \leq 0.99$

12) $1.23 < \dfrac{2.9 - 1.3x}{1.44} < 1.89$

13) Solve for σ, assuming $\sigma > 0$: $64 < \sigma^2 < 121$

14) Solve for σ, assuming $\sigma > 0$: $1.254 < \sigma^2 < 2.344$

15) Solve for $b - a$: $-0.34 < a - b < 0.44$

16) Solve for $x_1 - x_2$: $-1.15 < x_2 - x_1 < 2.67$

Answers to Exercises

1.1 page 3

1)   a) 2.9   b) −10.3   c) 153.8   d) 0.1   e) 0.0

2)   a) 4.73   b) −0.03   c) 1.45   d) −143.14   3)   78.8   4)   0.36

1.2 page 7

1)   −19   2) 46   3) −46   4) 13.31   5) 9.3, No.   6) 3.82

7) $4xy + 2y - x$ 8) 0   9) $3xy - 6x - xy^2 + 2y$ 10) $2xz - 3yz$

1.3 page 13

1) $\dfrac{1}{2}$ 2) $\dfrac{1}{4}$ 3) $\dfrac{3}{5}$ 4) $\dfrac{1}{6}$ 5) $-\dfrac{28}{51}$ 6) $\dfrac{9}{14}$

7)   0.965   8)   −5.69   9) $\dfrac{21y + 14}{3y}$ 10) $\dfrac{x - xy + 2 - 2y}{(1 + y)x}$

11) $\dfrac{y}{y - 2}$ 12) $\dfrac{w}{xy}$ 13) $\dfrac{(x + y)^2}{x}$ 14) $\dfrac{1}{y(x - y)}$

1.4 page 18

1) $\dfrac{4}{5}$ 2) $\dfrac{5}{6}$ 3) $\dfrac{1}{6}$ 4) $\dfrac{7}{12}$ 5) $\dfrac{11}{8}$ 6) $\dfrac{7}{30}$

7) $\dfrac{5}{16}$ 8) $\dfrac{37}{30}$ 9) $-\dfrac{32}{18} = -\dfrac{16}{9}$ 10) $\dfrac{22}{45}$ 11) $-\dfrac{25}{66}$

12) $\dfrac{29}{42}$ 13)   169.49   14)   27.1547   15)   0.44   16)   −3.153

17) $\dfrac{x+y}{xy}$

18) $\dfrac{x-y}{xy}$

19) $\dfrac{4yz-2xz+xy}{xyz}$

20) $\dfrac{yz-xz-z+xy-2y}{xyz}$

21) $\dfrac{x(z-xy)}{y(x-z)}$

22) $\dfrac{w-st}{s-2tw}$

1.5 page 23

1) 50   2) .2343   3) 1304.4   4) 52.8% female, 47.2% male

5) 36.8%   6) a) 2.5%   b) 6.3%   c) 15.2%   d) 34.2%   e) 41.8%

7) a) $12.74 b) $1124.25   8) $52,500.00

9) a) $8000 b) $11,000 c) $3000 d) 37.5%

1.6 page 29

1)

2) a) $[-1,3]$

   b) $(-1,3]$

   c) $[-3,1)$

   d) $[-3,4]$

   e) $(-\dfrac{1}{2},\sqrt{2}]$

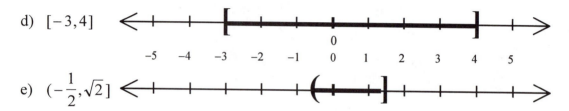

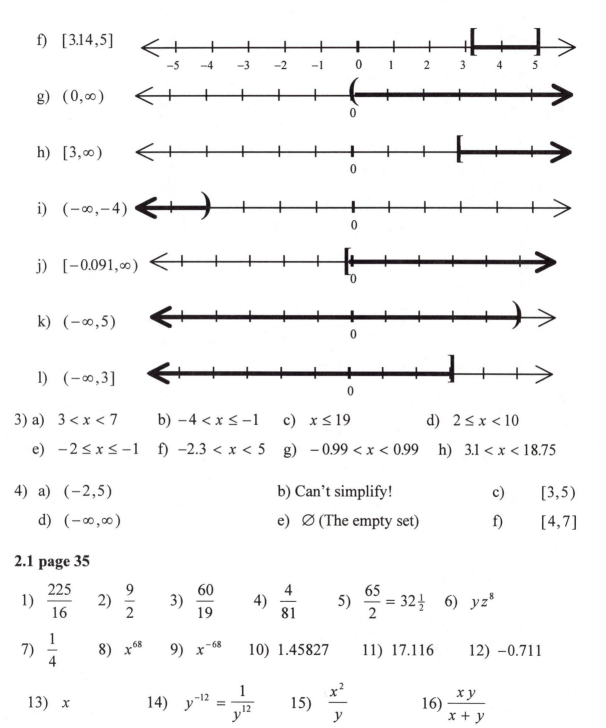

f) $[3.14,5]$

g) $(0,\infty)$

h) $[3,\infty)$

i) $(-\infty,-4)$

j) $[-0.091,\infty)$

k) $(-\infty,5)$

l) $(-\infty,3]$

3) a) $3 < x < 7$ b) $-4 < x \le -1$ c) $x \le 19$ d) $2 \le x < 10$

e) $-2 \le x \le -1$ f) $-2.3 < x < 5$ g) $-0.99 < x < 0.99$ h) $3.1 < x < 18.75$

4) a) $(-2,5)$ b) Can't simplify!   c) $[3,5)$

d) $(-\infty,\infty)$ e) $\varnothing$ (The empty set)   f) $[4,7]$

2.1 page 35

1) $\dfrac{225}{16}$ 2) $\dfrac{9}{2}$ 3) $\dfrac{60}{19}$ 4) $\dfrac{4}{81}$ 5) $\dfrac{65}{2} = 32\frac{1}{2}$ 6) yz^8

7) $\dfrac{1}{4}$ 8) x^{68} 9) x^{-68} 10) 1.45827 11) 17.116 12) -0.711

13) x 14) $y^{-12} = \dfrac{1}{y^{12}}$ 15) $\dfrac{x^2}{y}$ 16) $\dfrac{xy}{x+y}$

2.2 page 40

1) a) 3.83×10^5 b) -7.24×10^{-4} c) 3.00   d) 2.00×10^2

2) a) 3417.8   b) -109.15 c) 0.0000054378

    d) -0.0476 e) 80,000,100,000

3) a) 9.48×10^7 b) -3.09×10^5 c) 3.50×10^{-100}

    d) 7.66×10^{-2} e) -1.68×10^{-8} f) 5.57×10^{-2}

4) a) \$8017.84   b) \$3714.87 48.6% increase

2.3 page 44

1) 12   2) $\dfrac{1}{3}$ 3) $\dfrac{2}{7}$ 4) 3.87

5) $\sqrt{-11}$ Doesn't exist as a real number.

6) 1.34   7) 4.05   8) 13   9) 1.82

10) a) No! Let $a = b = 1$, $\sqrt{2}$ is certainly not equal to 2.

    b) Yes! If $a = 0$ and $b = 1$, then $\sqrt{a^2 + b^2} = \sqrt{1} = 1 = a + b$

11) No! Let $x = y = \dfrac{5}{\sqrt{2}}$, then $x^2 + y^2 = 25$ but $x + y = \dfrac{10}{\sqrt{2}}$.

3.1 page 51

1) 9   2) 11½   3) $\dfrac{15}{16}$

4) 129   5) \$7.50   6) Team A

7) 2.79 hours per week   8) 70.3, 13.73   9) 149.9, 35.64

3.2 page 55

1) 82.2   2) 73.6   3) 4.967%   4) 18.1

4.1 page 59

1) a) 12   b) $\dfrac{1}{56}$ c) 21   2) a) 362,880   b) 479,001,600

2) c) Calculator Answer: 3.04140932E64 (Approximately!)

   d) Too big to calculate on most calculators!

3) 6   4) 1

4.2 page 66

1) a) 1680   b) 70   2) a) 72   b) 9   c) 362,880   d) 362,880

3) a) 715   b) 13   c) 1   d) 715   4) 0   5) 27,216

6) 1.6661667×10^{11} 7) 1.5207×10^{64} 8) They are equal.

5.1 page 79

1) Number of Families

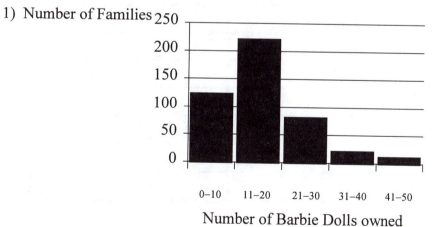

Number of Barbie Dolls owned

2) Using the following window parameters, gets the following histogram.

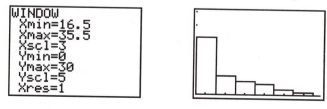

3)   a) 540   b) 580 on May 5th   c) 320 on May 18th   d) 487

e)

Week of April 19	Volume
Monday	380
Tuesday	518
Wednesday	480
Thursday	480
Friday	400

4)   a) $C = 6 + 1.5n$

b)

Number of Toppings	Cost of a Large Pie
0	$ 6.00
1	$ 7.50
2	$ 9.00
3	$ 10.50
4	$ 12.00
5	$ 13.50

c)

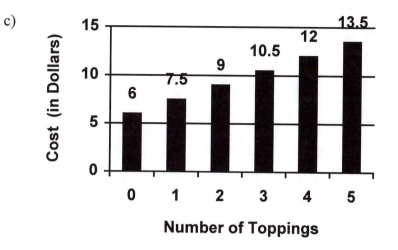

5)   Using the Stat Zoom we get the scatter diagram:

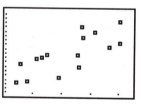

where the window parameters are:

```
WINDOW
 Xmin=60.22
 Xmax=73.18
 Xscl=3
 Ymin=113.93
 Ymax=209.07
 Yscl=5
 Xres=1
```

6.1 page 91

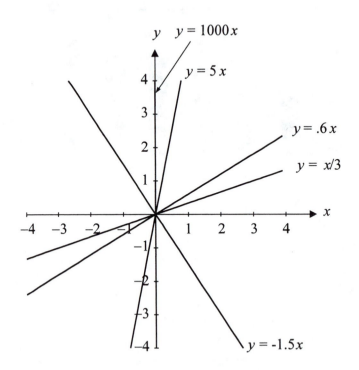

Believe it or not the graph for $y = 1000\,x$ is really plotted on this graph, but because of the scale you can't really see it. If we scale this as in the figure on the next page you see this line but lose the others.

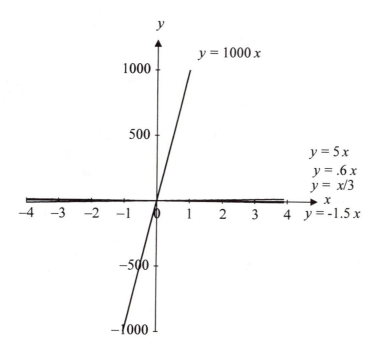

6)   Slope: 5.23, Intercept (0,4.56)

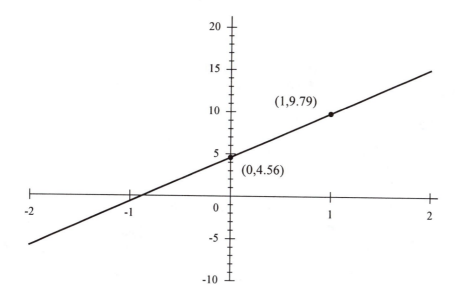

7) Slope: −15.47, Intercept (0,1.28)

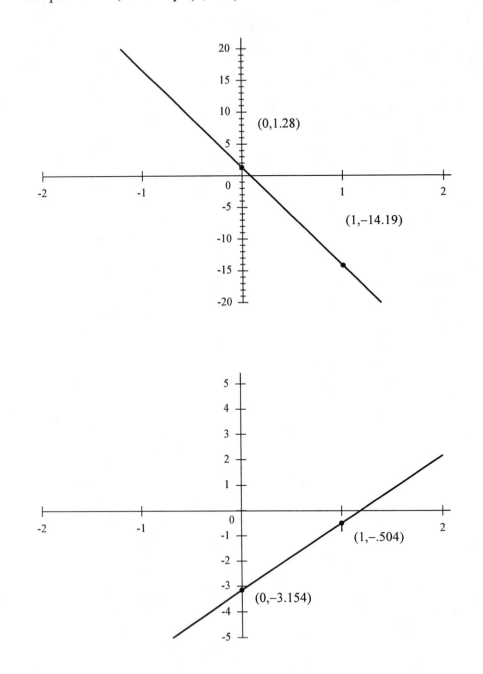

8)

9)

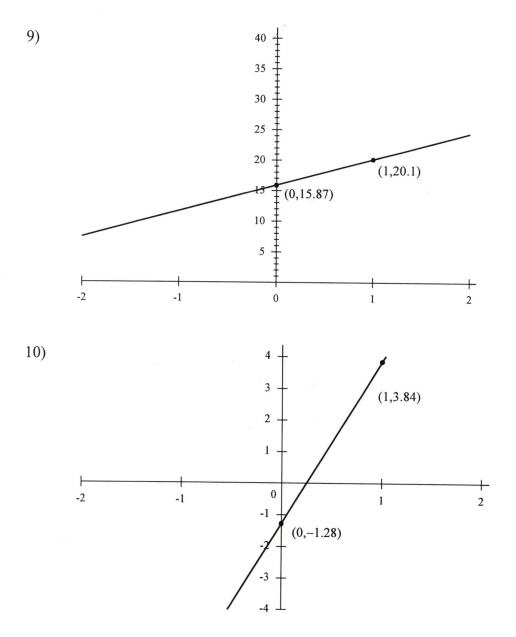

(1,20.1)

(0,15.87)

10)

(1,3.84)

(0,−1.28)

11)

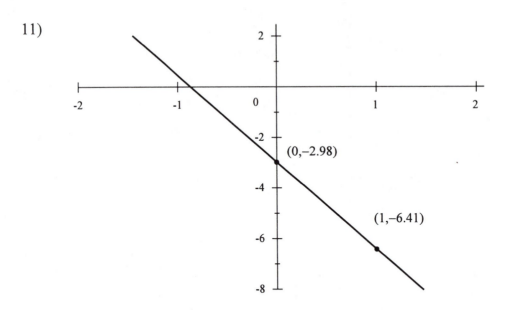

(0,−2.98)

(1,−6.41)

12)   Using the following window:

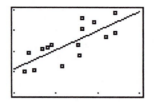

```
WINDOW
 Xmin=60
 Xmax=75
 Xscl=5
 Ymin=100
 Ymax=200
 Yscl=25
 Xres=1
```

we get the graph:

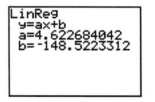

where regression parameters are:

```
LinReg
 y=ax+b
 a=4.622684042
 b=-148.5223312
```

7.1 page 102

1) $\dfrac{4}{3}$

2) 84

3) $\dfrac{33}{40}$

4) $x = \sigma z + \mu$

5) 256

6) $n = \left[\dfrac{t_{\alpha/2} \cdot s}{E}\right]^2$

7) $n = \dfrac{z_{\alpha/2}^2 \, p \, q}{E^2}$

8) $\dfrac{1}{5}$

9) $x = -\dfrac{z^2}{3y^2}$

10) $y = -\dfrac{3z + 1}{2z}$

7.2 page 107

1) $-1, -4$

2) $-2, 4$

3) repeated root, -3

4) $\dfrac{1}{2}, -1$

5) Repeated root; 4

6) $-4 \pm 2\sqrt{3}$

7) $\pm \dfrac{2}{\sqrt{3}} = \pm \dfrac{2\sqrt{3}}{3}$

8) $1, -4$

9) $\dfrac{-7 \pm \sqrt{33}}{2}$

10) $\dfrac{2}{5}, -1$

7.3 page 114

1) $x > 12$

2) $x < \dfrac{7}{3}$

3) $x \le -2$

4) $x \ge -3$

5) $x > 6.938461538$

6) $x > 1.434883721$

7) $\dfrac{4}{5} < x < \dfrac{29}{5}$

8) $\dfrac{-11}{4} \le x < \dfrac{1}{4}$

9) $-2 \le x < -1$

10) $2.85 < x < 3.55$

11) $-0.581130435 \le x \le -0.288434783$

12) $0.137230769 < x < 0.868307692$

13) $8 < \sigma < 11$

14) $1.1198 < \sigma < 1.531$

15) $-0.44 < b - a < 0.34$.

16) $-2.67 < x_1 - x_2 < 1.15$.

Index